Unstoppable
Global Warming

Unstoppable Global Warming

Every 1,500 Years

S. Fred Singer and Dennis T. Avery

ROWMAN & LITTLEFIELD PUBLISHERS, INC.
Lanham • Boulder • New York • Toronto • Plymouth, UK

ROWMAN & LITTLEFIELD PUBLISHERS, INC.

Published in the United States of America
by Rowman & Littlefield Publishers, Inc.
A wholly owned subsidary of The Rowman & Littlefield Publishing Group, Inc.
4501 Forbes Boulevard, Suite 200, Lanham, Maryland 20706
www.rowmanlittlefield.com

Estover Road
Plymouth PL6 7PY
United Kingdom

British Library Cataloguing in Publication Information Available

Library of Congress Cataloging-in-Publication Data

Singer, S. Fred (Siegfried Fred), 1924–
  Unstoppable global warming : every 1500 years / S. Fred Singer and Dennis T. Avery.
    p. cm.
  Includes bibliographical references and index.
  ISBN-13: 978-0-7425-5116-9 (alk. paper)
  ISBN-10: 0-7425-5116-4 (alk. paper)
  ISBN-13: 978-0-7425-5117-6 (pbk. : alk. paper)
  ISBN-10: 0-7425-5117-2 (pbk. : alk. paper)
  1. Global warming. 2. Global temperature changes. 3. Greenhouse effect, Atmospheric.
I. Avery, Dennis T. II. Title.
QC981.8.G56S553 2006
551.6—dc22   2006009308

Printed in the United States of America

∞™ The paper used in this publication meets the minimum requirements of American
National Standard for Information Sciences—Permanence of Paper for Printed Library
Materials, ANSI/NISO Z39.48-1992.

DEDICATION

This book is dedicated to those thousands of highly qualified research scientists who have documented physical evidence of the 1,500-year climate cycle from over the entire globe. Hundreds of their studies endorse the reality of this cycle. They have gone, literally, to the ends of the Earth—often in harsh conditions, climbing tall mountains and huge glaciers, sailing frigid seas, sitting cramped at consoles in long-range aircraft, digging painstakingly in remote sites for pollen and human artifacts—to help us understand our planet's climate changes. They have done it with virtually no encouragement from the press and the public. Too often, they have had to work in the face of hostility from their colleagues in the climate modeling community who are "selling" CO_2. They have done it in a sincere effort to help produce public policies that will maximize both human and ecological well-being. We applaud them and their work.

The three scientists, Willi Dansgaard of Denmark, Hans Oeschger of Switzerland, and Claude Lorius of France, who led the discovery of the 1,500-year cycle, were jointly awarded the Tyler Prize (the "environmental Nobel") in 1996. However, their award citations say nary a word about the climate cycle they discovered, nor anything about its predictive power to forecast moderate climate changes. The public has remained virtually unaware that the 1,500-year cycle offers the only explanation for the modern warming that is supported by physical evidence.

Contents

List of Figures

Prologue

Living and Dying by Greenland's 1,500-Year Climate Cycle

When Eric the Red led Norse families to settle on Greenland toward the end of the tenth century, he had no idea that he and his descendants were about to demonstrate the Earth's long, moderate climate cycle as dramatically as it would ever be done.

Sailing their longships west from Iceland, the Vikings had been pleased to find a huge new uninhabited island, its shores covered with green grass for their cattle and sheep, surrounded by ice-free waters where codfish and seals abounded. They could grow vegetables for their families and hay to feed their animals through the winter. There was no timber but they could ship dried fish, sealskins, and tough rope made from walrus hide to other Norse ports to trade for what they needed. The colony thrived, growing by the year 1100 to three thousand people, with twelve churches and its own bishop. The initial settlement split into two: one on the southwestern coast and one further north also on the west coast.

They did not realize that they were benefiting from the Medieval Warming, a major climate shift—which lasted for approximately four hundred years—that made Northern Europe about 2° Celsius warmer than it had been previously. Nor did they realize that after the warming ended, their grassy domain was doomed to five hundred years—the Little Ice Age—of icy temperatures unmoderated by the Gulf Stream that warmed the Norse settlements in Norway and Iceland.

As the Little Ice Age progressed, the colony was increasingly hard-pressed to survive. The pack ice moved closer to Greenland. Supply ships had to take a more southerly route to avoid the deadly ice. Less and less hay could be harvested in the shorter, cooler summers to last the livestock through longer and colder winters. The storms got worse.

By 1350, glaciers had crushed the northern Greenland settlement. The last supply ship got through to the southern settlements in 1410; then they were cut off. There was fighting with Inuit hunters pushed south by the encroaching ice to compete with the Norse for the seals. The codfish followed the warm water south, away from the colonies. The skeletons in the graveyards show the people growing shorter—indicating poor nutrition.

We do not know whether the last colonists died at the hands of the Inuit, or froze or starved to death, or even just when they died. We do know they had eaten the last of their milk cows—a desperate act for dairy farmers. Measuring the change in oxygen isotopes of Greenland Norse skeletons' tooth enamel indicates a 1.5-degree Celsius drop in average temperatures between the years 1100 and 1400.

Denmark would not recolonize Greenland until 1721, when the Little Ice Age was losing its grip on the huge island.

Today, 150 years into the Modern Warming, Greenland has 50,000 people and 20,000 sheep. Most of the people earn their living by catching shrimp and fish, though there is a short summer season for hardy tourists. The ruins of the Norse cathedral and the bishop's palace have been partially restored as a beautiful memorial to Greenland's Norse era.

Greenland is likely to become even more popular as a tourist attraction for as long as the Modern Warming lasts—probably at least several hundred more years. But the ice cores and seabed sediments tell us of six hundred natural 1,500-year climate cycles over the past one million years. The cycle will eventually shift again and Greenland will once more descend into ice and hardship.

The only significant difference between the fifteenth century's Little Ice Age and the next is likely to be in the human technology available to cope with it. The twenty-fifth century's Greenlanders will have better insulation and bigger, safer fishing vessels than did their ancestors, plus satellite communications and powerful, ice-breaking supply ships.

Earth's Climate Timeline

ANCIENT CLIMATE HISTORY

4.5 billion years ago: Earth created.

3.8 billion years ago: Single-cell life forms created.

1.9 billion years ago: Nitrogen-dominated atmosphere replaced by oxygen-rich atmosphere.

540 million years ago: Earth apparently turned from "ice ball" into a warm and humid world. Simple microbes in the ocean transform into thousands of new species, founding the evolution of today's life forms.

350 to 250 million years ago: Ice sheets reappear at higher Earth latitudes. There is only one large continental landmass, Gondwana, in the southern hemisphere.

300 million years ago: Earth uniformly hot and humid, land area dominated by swamps and rain forest.

250 million years ago: Permian period, hot and so dry that huge salt deposits are formed from ocean evaporation.

100 million years ago: Age of Dinosaurs. The Earth is still warm, again humid, and globally uniform. Continents are slowly separating through plate tectonics.

60 million years ago: World in a "greenhouse" condition without ocean circulation patterns or polar regions.

30 million years ago: Antarctica separates from South America and the Southern Ocean is born. Shortly after, glaciers start to expand in Antarctica.

5 million years ago: Modern climate begins to develop as separation of continents produces the great oceanic temperature conveyor in the North Atlantic. Northern Hemisphere cooling leads to more glaciers and ice sheets.

2 million years ago: Cycles in Earth's relation to the sun produce alternating Ice Ages (90,000 to 100,000 years) and "interglacials" (10,000 to 20,000 years). The onset of the glacial period is often slow but ends abruptly at the transition to the warm period. The average global temperature changes 5 to 7 degrees Celsius at this transition but may rise as much as 10 to 15 degrees Celsius over a time span of less than seventy-five years at higher latitudes.[1]

THE MORE RECENT CLIMATE TIMELINE

130,000 to 110,000 years ago: Eemian interglacial, warm.

110,000 years ago: Fairly sudden shift to much-colder-than-present glacial conditions, over perhaps four hundred years—or even less. Northern forests retreat south, ice sheets begin to take over much of Northern Hemisphere. Trees give way to grass, and then to deserts, as more water is frozen in ice sheets instead of falling as rain on vegetation.

60,000 to 55,000 years ago: In-between phase, partial melting of glaciers.

30,000 years ago: Last Ice Age reached its coldest point 21,000–17,000 years ago. Deserts and semi-deserts take over much of the global land area. Sea levels four hundred feet lower than today.

14,000 years ago: Sudden warming, raising Earth temperatures to roughly present levels. Forests began to spread and the ice sheets to retreat. Sea levels begin to rise.

12,500 years ago: The Younger Dryas. After only 1,500 years of recovery from the Ice Age, the Earth was suddenly plunged back into a new, short-lived ice age. The dramatic cooling seems to have occurred within one hundred years or less. Another 1,000 years or so of ice age followed before another sudden shift back to climate warming.

11,500 years ago: The present interglacial period, the Holocene. The planet warmed from ice age to nearly present world temperatures in less than one hundred years. Half of the warming may have occurred in fifteen years. Ice sheets melted, sea levels rose again, and forests expanded. Trees replaced grass and grass replaced deserts.

9,000 to 5,000 years ago: Climate Optimum, warmer and wetter than the Earth's present climate. The Saharan and Arabian deserts became wetter, supporting hunting, herding, and some agriculture. The climate may have been "punctuated" by a cold, dry phase 8,200 years ago, with Africa drier than before.

2,600 years ago: Cooling event with relatively wet conditions in many parts of the world.

The 1,500-Year Cycle: Through at least the last one million years, and the recent ice ages, a 1,500-year warm–cold cycle has been superimposed over the longer, stronger ice ages and interglacial phases. In the North Atlantic, the temperature changes, from peak to trough, of these "Dansgaard-Oeschger cycles" has been about 4°C. The shift into cold phases has often been very abrupt.

RECENT EARTH CLIMATE HISTORY

600 to 200 B.C.: Unnamed cold period that preceded the Roman Warming.
200 B.C. to about A.D. 600: Roman Warming.
600 to 900: Dark Ages cold period.
900 to 1300: Medieval Warming or Little Climate Optimum.
1300 to 1850: Little Ice Age (two-stage).

MODERN CLIMATE HISTORY

1850 to 1940: Warming, especially between 1920 and 1940.
1940 to 1975: Cooling trend.
1976 to 1978: Sudden warming spurt.
1979 to present: A large disparity between surface thermometers, which show a fairly strong warming, and the independent temperature readings of satellites and balloons, which show little warming trend.

NOTE

1. Mikkelsen, N., and A. Kuijpers, *The Climate System and Climate Variations*, "Natural Climate Variations in a Geological Perspective," Geological Survey of Denmark and Greenland, 2001.

Chapter One

Is Humanity Losing the Global Warming Debate?

The Earth is warming but physical evidence from around the world tells us that human-emitted CO_2 (carbon dioxide) has played only a minor role in it. Instead, the mild warming seems to be part of a natural 1,500-year climate cycle (plus or minus 500 years) that goes back at least one million years.

The cycle has been too long and too moderate for primitive peoples lacking thermometers to recount in their oral histories. But written evidence of climatic change does exist. The Romans had recorded a warming from about 200 B.C. to A.D. 600, registered mainly in the northward advance of grape growing in both Italy and Britain. Histories from both Europe and Asia tell us there was a Medieval Warming that lasted from about 900 to 1300; this period was also known as the Medieval Climate Optimum because of its mild winters, stable seasons, and lack of severe storms. Human histories also record the Little Ice Age, which lasted from about 1300 to 1850. But people thought each of these climatic shifts was a distinct event and not part of a continuing pattern.

This began to change in 1984 when Willi Dansgaard of Denmark and Hans Oeschger of Switzerland published their analysis of the oxygen isotopes in the first ice cores extracted from Greenland.[1] These cores provided 250,000 years of the Earth's climate history in one set of "documents." The scientists compared the ratio of "heavy" oxygen-18 isotopes to the "lighter" oxygen-16 isotopes, which indicated the temperature at the time the snow had fallen. They expected to find evidence of the known 90,000-year ice ages and the mild interglacial periods recorded in the ice, and they did. However, they did not expect to find anything in between. To their surprise, they found a clear cycle—moderate, albeit abrupt—occurring about every 2,550 years running persistently through both. (This period would soon be reassessed at 1,500 years, plus or minus 500 years.)

By the mid-1980s, however, the First World had already convinced itself of the Greenhouse Theory and believed that puny human industries had grown powerful enough to change the planet's climate. There was little media interest in the frozen findings of obscure, parka-clad Ph.D.s in far-off Greenland.

A wealth of other evidence has emerged since 1984, however, corroborating Dansgaard and Oeschger's natural 1,500-year climate cycle:

- An ice core from the Antarctic's Vostok Glacier—at the other end of the world from Iceland—was brought up in 1987 and showed the same 1,500-year climate cycle throughout its 400,000-year length.
- The ice-core findings correlate with known advances and retreats in the glaciers of the Arctic, Europe, Asia, North America, Latin America, New Zealand, and the Antarctic.
- The 1,500-year cycle has been revealed in seabed sediment cores brought up from the floors of such far-flung waters as the North Atlantic Ocean and the Sargasso Sea, the South Atlantic Ocean and the Arabian Sea.
- Cave stalagmites from Ireland and Germany in the Northern Hemisphere to South Africa and New Zealand in the Southern Hemisphere show evidence of the Modern Warming, the Little Ice Age, the Medieval Warming, the Dark Ages, the Roman Warming, and the unnamed cold period before the Roman Warming.
- Fossilized pollen from across North America shows nine complete reorganizations of our trees and plants in the last 14,000 years, or one every 1,650 years.
- In both Europe and South America, archaeologists have evidence that prehistoric humans moved their homes and farms up mountainsides during the warming centuries and retreated back down during the cold ones.

The Earth continually warms and cools. The cycle is undeniable, ancient, often abrupt, and global. It is also unstoppable. Isotopes in the ice and sediment cores, ancient tree rings, and stalagmites tell us it is linked to small changes in the irradiance of the sun.

The temperature change is moderate. Temperatures at the latitude of New York and Paris moved about 2 degrees Celsius above the long-term mean during warmings, with increases of 3 degrees or more in the polar latitudes. During the cold phases of the cycle, temperatures dropped by similar amounts below the mean. Temperatures change little in lands at the equator, but rainfall often does.

The cycle shifts have occurred roughly on schedule whether CO_2 levels were high or low. Based on this 1,500-year cycle, the Earth is about 150 years into a moderate Modern Warming that will last for centuries longer. It will essentially restore the fine climate of the Medieval Climate Optimum.

The climate has been most stable during the warming phases. The "little ice ages" have been beset by more floods, droughts, famines, and storminess. Yet, despite all of this evidence, millions of well-educated people, many scientists, many respected organizations—even the national governments of major First World nations—are telling us that the Earth's current warming phase is caused by human-emitted CO_2 and deadly dangerous. They ask society to renounce most of its use of fossil fuel-generated energy and accept radical reductions in food production, health technologies, and standards of living to "save the planet."

We have missed the predictive power of the 1,500-year climate cycle.

Will the fear of dangerous global warming lead society to accept draconian restrictions on the use of fertilizers, cars, and air conditioners?

Will people give up the scientific and technological advances that have added thirty years to life expectancies all over the globe in the last century?

Massive human sacrifices would be required to meet the CO_2 stabilization goals of the Kyoto Protocol. The treaty's "introductory offer" is a tiny 5 percent reduction in fossil fuel emissions from 1990 levels, but that would do almost nothing to forestall greenhouse warming of the planet. Saving the planet from man-made global warming was supposed to wait on Kyoto's yet-unspecified second stage, scheduled to begin in 2012.

In 1995, one U.S. environmentalist assessed the outlook: "According to the [United Nations] Intergovernmental Panel on Climate Change, an immediate 60 to 80 percent reduction in emissions is necessary just to stabilize atmospheric concentrations of CO_2—the minimum scientifically defensible goal for any climate strategy. Less-developed nations, with their relatively low emissions, will inevitably increase their use of fossil fuels as they industrialize and their populations expand. Thus heavily polluting regions like the [United States] will have to reduce their emissions even more [than 60 to 80 percent] for the world as a whole to meet this goal."[2]

Humans use eighty million tons per year of nitrogen fertilizer to nourish their crops. The nitrogen is taken from the air (which is 78 percent N_2) through an industrial process generally fueled by natural gas. In 1900, before industrial nitrogen fertilizer, the world could support only 1.5 billion people, at a far lower standard of living, and was clearing huge tracts of forest to get more cropland.

Suppose the world went all-organic in its farming, gave up the man-made fertilizer, and cleared half of the world's remaining forests for more low-yield crops. It's reasonable to expect that half the world's wildlife species would be lost in the land clearing and one-fourth of the world's people would succumb to malnutrition. What if research then confirmed that the climate was warming due to the natural cycle instead of CO_2? Is that a no-regrets climate insurance policy?

What if the Kyoto treaty or some similar arrangement prevented the Third World from moving away from using wood for heating and cooking? How much additional forest would then be sacrificed for firewood in the developing countries over the next fifty years?

The stakes in the global warming debate are huge. Humanity and wildlife may both be losing the debate.

DOES A 1,500-YEAR CLIMATE CYCLE RULE OUR EARTH?

Greenhouse Warming Advocates Say:

"Nineteen ninety-nine was the most violent year in the modern history of weather. So was 1998. So was 1997. And 1996. . . . A nine-hundred-year-long cooling trend has been suddenly and decisively reversed in the past fifty years. . . . Scientists predicted that the Earth will shortly be warmer than it has been in millions of years. A climatological nightmare is upon us. It is almost certainly the most dangerous thing that has ever happened in our history."[3]

"Climate extremes would trigger meteorological chaos—raging hurricanes such as we have never seen, capable of killing millions of people; uncommonly long, record-breaking heat waves; and profound drought that could drive Africa and the entire Indian subcontinent over the edge into mass starvation."[4]

"From sweltering heat to rising sea levels, global warming's effects have already begun. . . . We know where most heat-trapping gases come from: power plants and vehicles. And we know how to limit their emissions."[5]

"Such policies like cutting energy use by more than 50 percent can contribute powerfully to the material salvation of the planet from mankind's greed and indifference."[6]

"No matter if the science of global warming is all phony . . . climate change [provides] the greatest opportunity to bring about justice and equality in the world."[7]

Reality-Based Skeptics Say:

"The study, appearing in the March 21 issue of the journal *Science*, analyzed ancient tree rings from 14 sites on three continents in the northern hemisphere

Holiday Inn®

holiday-inn.co.uk
+800 40 50 60

and concluded that temperatures in an era known as the Medieval Warm Period some 800 to 1,000 years ago closely matched the warming trend of the 20th century."[8]

"I want to encourage the committee to be suspicious of media reports in which weather extremes are given as proof of human-induced climate change. Weather extremes occur somewhere all the time. For example, in the year 2000 in the 48 coterminous states, the U.S. experienced the coldest combined November and December in 106 years. . . . The intensity and frequency of hurricanes have not increased. The intensity and frequency of tornados have not increased. . . . Droughts and wet spells have not statistically increased or decreased."[9]

"Hurricanes, brutal cold fronts and heat waves, ice storms and tornadoes, cycles of flood and drought, and earthquakes and volcanic eruptions are not unforeseeable interruptions of normality. Rather, these extremes are the way that the planet we live on does its business. Hurricanes, in some parts of the world, provide a third of the average annual rainfall. What we call "climate" is really an average of extremes of heat and cold, precipitation and drought. . . . [A]ll the evidence from paleoclimatology and geology suggests that over the long haul, the extremes we face will be substantially greater than even the strongest in our brief historical record."[10]

"[T]he number of major [Chinese] floods averaged fewer than four per century in the warm period of the ninth through eleventh centuries, while the average number was more than double that figure in the fourteenth through seventeenth centuries of the Mini Ice Age."[11]

The Purpose of This Book

The purpose of this book is to offer the relatively new but already convincing evidence of a moderate, irregular 1,500-year sun-driven cycle that governs most of the Earth's almost-constant climate fluctuations.

The Earth has recently been warming. This is beyond doubt. It has warmed slowly and erratically—for a total of about 0.8 degrees Celsius—since 1850. It had one surge of warming from 1850 to 1870 and another from 1920 to 1940. However, when we correct the thermometer records for the effects of growing urban heat islands and widespread intensification of land use, and for the recently documented cooling of the Antarctic continent over the past thirty years, overall world temperatures today are only modestly warmer than they were in 1940, despite a major increase in human CO_2 emissions.

The real question is not whether the Earth is warming but why and by how much.

We have a large faction of intensely interested persons who say the warming is man-made, and dangerous. They say it is driven by releases of greenhouse gases such as CO_2 from power plants and autos, and methane from rice paddies and cattle herds. The activists tell us that modern society will destroy the planet; that unless we radically change human energy production and consumption, the globe will become too warm for farming and the survival of wild species. They warn that the polar ice caps could melt, raising sea levels and flooding many of the world's most important cities and farming regions.

However, they don't have much evidence to support their position—only (1) the fact that the Earth is warming, (2) a theory that doesn't explain the warming of the past 150 years very well, and (3) some unverified computer models. Moreover, their credibility is seriously weakened by the fact that many of them have long believed modern technology should be discarded whether the Earth is warming too fast or not at all.

Many scientists—though by no means all—agree that increased CO_2 emissions could be dangerous. However, polls of climate-qualified scientist show that many doubt the scary predictions of the global computer models. This book cites the work of many hundreds of researchers, authors, and coauthors whose work testifies to the 1,500-year cycle. There is no "scientific consensus," as global warming advocates often claim. Nor is consensus important to science. Galileo may have been the only man of his day who believed the Earth revolved around the sun, but he was right! Science is the process of developing theories and *testing them against observations* until they are proven true or false.

If we can find proof, not just that the Earth is warming, but that it is warming to dangerous levels due to human-emitted greenhouse gases, public policy will then have to evaluate such potential remedies as banning autos and air conditioners. So far, we have no such evidence.

If the warming is natural and unstoppable, then public policy must focus instead on adaptations—such as more efficient air conditioning and building dikes around low-lying areas like Bangladesh. We have the warming. Now we must ascertain its cause.

THE STORY OF THE 1,500-YEAR CLIMATE CYCLE

Carefully retrieved ice cores now reveal 900,000 years of the planet's climate changes. The information in the ice layers is being supplemented by such scientific breakthroughs as solar-monitoring satellites and mass spectrometer measurements of oxygen and carbon isotopes. These are allowing the Earth itself to tell us about its own climate history.

The ice cores say the newly understood 1,500-year cycle has dominated Earth's climate during the 11,000 years since the last ice age. Moreover, its fingerprints are being found all over the world, stretching through previous ice ages and interglacials.

We'll cover the evidence that past climate cycles such as the Medieval Warming and the Little Ice Age were truly global, not just Europe-only events as some man-made warming advocates have suggested. Our search for evidence will carry us from European castles to Chinese orange groves to Japanese cherry blossom viewings, from Saharan lakes to Andean glaciers and a South African cave.

No single climate proxy is totally equal to having thermometer records from the Middle Ages or ancient Rome. Tree rings reflect not only temperature and sunlight but also other factors that affect the tree, including rainfall, the number of competing trees, tree insects, and diseases. Borehole temperature signals become fainter as they go deeper.

Individually, each piece of proxy evidence could be questioned. That's why we'll examine a broad array of them, widespread in their geography, and almost dizzying in their variety.

We'll examine lots of tree rings, ice cores, and seabed sediments because they're the most important long-record proxies and among the most accurate, if properly treated. We'll also look at peat bogs full of antique organic residues; stalagmites from caves where varying amounts of moisture and minerals from the surface have dripped for century after century; coral reefs whose tiny creators have left behind clues to the sea temperatures while they were working; and ancient iron dust that betrays huge droughts.

We'll check out the reasons why many people today fear global warming. This fear of warm weather comes in dramatic contrast to our forebears—who loved warming climates and hated the mini-ice ages.

After all, it was during the cooling Dark Ages that the Roman and Mayan empires collapsed, after they had thrived during a warming that was hotter than it is today. And, it was during the cold of the Little Ice Age that Europe had its worst-ever floods and famines.

EARTH'S CLIMATE LINK TO THE SUN

Humans have known about a link between Earth's climate cycles and solar variability for more than four hundred years because of sunspot variations.

Most dramatically, the Maunder Sunspot Minimum occurred from 1640 to 1710, when there were virtually no sunspots at all for some seventy years. That marked the sun's weakest recent moment—and that was the very coldest point in the Little Ice Age. Observers have also known that sunspot cycles that lasted longer than the average eleven years (the variation is eight to fourteen years) produce warmer temperatures on Earth. What our ancestors didn't know was how the solar-Earth climate link functioned.

Until the age of satellites, we didn't even know that tiny cycles of variation in the sun's irradiance existed. Until recently, scientists spoke of the "solar constant." Now, measuring from outside the clouds and gases of the Earth's atmosphere, we find that the sun's intensity varies by fractions of a percentage point.

Until 2001, the global warming debate was what a lawyer would call a "he-said/she-said" controversy. The science for man-made global warming looked weak. The science for solar-driven climate cycles on Earth looked more plausible, but was inconclusive.

But on 16 November 2001, the journal *Science* published a report on elegant research, done by unimpeachable scientists, giving us the Earth's climate history for the past 32,000 years—along with our climate's linkage to the sun. The late Gerard Bond and a team from the Columbia University's Lamont-Doherty Earth Observatory published "Persistent Solar Influence on North Atlantic Climate during the Holocene."[12]

Science's Richard Kerr wrote:

Paleo-oceanographer Gerard Bond and his colleagues report that the climate of the northern North Atlantic has warmed and cooled nine times in the past 12,000 years in step with the waxing and waning of the sun.

"It really looks like the sun has mattered to climate," says glaciologist Richard Alley of Pennsylvania State University. . . . "The Bond et al., data are sufficiently convincing that [solar variability] is now the leading hypothesis to explain the roughly 1,500-year oscillation of the climate seen since the last ice age, including the Little Ice Age of the 17th century," says Alley.[13]

Bond's sun-climate correlation rests on a rare combination of long, continuous, and highly detailed records of both climate change and solar activity. The climate record is Bond's well-known and laborious accounting of the microscopic bits of rock dropped by melting icebergs onto the floor of the northern North Atlantic over thousands of years.

Bond and his team found that the amounts of debris increased in abundance every 1,500 years (give or take 500) as the ice was carried farther out into a temporarily colder Atlantic. During the last Ice Age's coldest spells, huge amounts of ice were carried clear across the Atlantic's polar region and south as far as Ireland.

Bond's linkages to solar activity are the carbon-14 isotopes in tree rings and beryllium-10 isotopes in the Greenland ice cores. The carbon and beryllium isotopes are linked to solar activity through the intensity of solar-modulated cosmic rays, and are produced when cosmic rays strike the upper atmosphere. (Oxygen isotopes found in ice cores reveal past temperatures.)

The startling element of Bond's results is the close correlation found when the global temperatures and solar-strength records are laid next to each other.

How Does the Sun Change Our Climate?

How does a tiny change in the sun's irradiance make a big difference in the Earth's climate? First of all, we know that the linkage exists and is powerful, which puts the solar cycle far ahead of the Greenhouse Theory.

We now have strong, new scientific evidence of how the linkage works. The key amplifier is cosmic rays. The sun sends out a "solar wind" that protects the Earth from some of the cosmic rays bombarding the rest of the universe. When the sun is weak, however, more of the cosmic rays get through to the Earth's atmosphere. There, they ionize air molecules and create cloud nuclei. These nuclei then produce low, wet clouds that reflect solar radiation back into outer space. This cools the Earth.

Researchers using neutron chambers to measure the cosmic rays have recently found that changes in Earth's cosmic ray levels are correlated with the number and size of such cooling clouds.

The second amplifier is ozone chemistry in the atmosphere. When the sun is more active, more of its ultraviolet rays hit the Earth's atmosphere, shattering more oxygen (O_2) molecules—some of which reform into ozone (O_3). The additional ozone molecules absorb more of the near-UV radiation from the sun, increasing temperatures in the atmosphere. Computer models indicate that a 0.1 percent change in the sun's radiation could cause a 2 percent change in the Earth's ozone concentration, affecting atmospheric heat and circulation.

INTERGOVERNMENTAL PANEL ON CLIMATE CHANGE CHARGES MAN-MADE WARMING

What about the claim of the United Nations' Intergovernmental Panel on Climate Change (IPCC): that they've found a "human fingerprint" in the current global warming?

That statement was inserted in the executive summary of the IPCC's 1996 report for political, not scientific, reasons. Then the "science volume" was edited to take out five different statements—all of which had been approved by the panel's scientific consultants—specifically saying no such "human fingerprint" had been found.

The author of the IPCC science chapter, a U.S. government employee, publicly admitted making the scientifically indefensible "back room" changes. He was under pressure from top U.S. government officials to do so.[14]

THE FAILURES OF THE GREENHOUSE THEORY

Let's quickly review the shortcomings of the Greenhouse Theory for explaining known realities.

First, and most obvious, CO_2 changes do not account for the highly variable climate we know the Earth has recently had, including the Roman Warming, the Dark Ages, the Medieval Warming, and the Little Ice Age. However, these variations fit into the 1,500-year cycle very well.

Second, the Greenhouse Theory does not explain recent temperature changes. Most of the current warming occurred *before 1940*, before there was much human-generated CO_2 in the air. After 1940, temperatures declined until 1975 or so, despite a huge surge in industrial CO_2 during that period. These events run *counter* to the CO_2 theory, but they are in accord with the 1,500-year cycle.

Third, the early and supposedly most powerful increases in atmospheric CO_2 have not produced the frightening planetary overheating that the theory and climate models told us to expect. We must discount *future* increments of CO_2 in the atmosphere, because each increment of CO_2 increase produces less warming than the unit before it. The amounts of CO_2 already added to the atmosphere must already have "used up" much—and perhaps most—of CO_2's forcing capability.

Fourth, we must discount the "official" temperature record to reflect the increased size and intensity of today's urban heat islands, where most of the official thermometers are located. We must take account of the changes in rural land use (forests cleared for farming and pastures, more intensive row-crop and irrigated farming) that affect soil moisture and temperatures. When meteorological experts reconstructed U.S. official temperatures "without cities and crops"—using more accurate data from satellites and high-altitude weather balloons—*about half of the recent "official" warming disappeared.*

Fifth, the Earth's surface thermometers have recently warmed faster than the temperature readings in the lower atmosphere up to 30,000 feet. Yet the Greenhouse Theory says that CO_2 will warm the lower atmosphere first, and then the atmospheric heat will radiate to the Earth's surface. *This is not happening.*

Figure 1.1 shows the very moderate trend in the satellite readings over the past two decades, totaling 0.125 degrees Celsius per decade. The short-term temperature spike in 1998 was one of the strongest El Niño events in recent centuries, but its effect quickly dissipated, as always happens with El Niños.

A reconstruction of weather-balloon temperature readings at two meters above the Earth's surface (1979–1996) shows a trend increase of only 0.015 degree Celsius per decade.[15] Nor can we project even that slow increase over the coming centuries, since the 1,500-year cycles have often achieved half of their total warming in their first few decades, followed by erratic warmings and coolings like those we've recorded since 1920.

Sixth, CO_2 for at least 240,000 years has been a *lagging indicator* of global warming, not a causal factor. Within the last 15 years, the ice cores have revealed that temperatures and CO_2 levels have tracked closely together during the warmings after each of Earth's last three ice age glaciations. However, the CO_2 changes have lagged about 800 years behind the temperature changes. *Global warming has produced more CO_2, rather than more CO_2 producing global warming.* This accords with the reality that the oceans hold the vast majority of the planet's carbon, and the laws of physics let cold oceans hold more CO_2 gas than warm oceans.

Seventh, the Greenhouse Theory predicts that CO_2-driven warming of the Earth's surface will start, and be strongest, in the North and South Polar

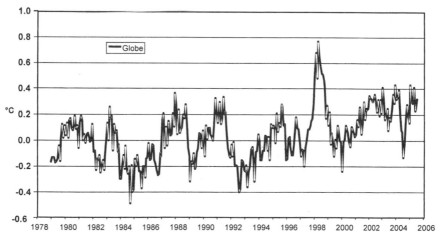

Figure 1.1.　Satellite Temperature Record 1979–2004, trending up at a modest 0.125 degrees Celsius per decade. A strong El Niño occurred in 1998.
Compiled by John Christy, University of Alabama–Huntsville.

regions. *This is not happening either.* A broadly scattered set of meteorological stations and ocean buoys show that temperature readings in the Arctic, Greenland, and the seas around them are colder today than in the 1930s. Alaska has been warming, but researchers say this is due to the recent warming of the Pacific Decadal Oscillation (PDO), not a broader Arctic warming pattern. The twenty to thirty year cycle of the PDO seems to have recently reversed again, so Alaska may now cool with the rest of the Arctic.

In the Antarctic, only the thin finger of the Antarctic Peninsula, which juts up toward Argentina (and the equator) has been warming. Temperatures over the other 98 percent of the Antarctic continent have been declining slowly since the 1960s, according to a broad array of Antarctic surface stations and satellite measurements.

Eighth, the scary predictions of planetary overheating require that the warming effect of additional CO_2 be amplified by increased water vapor in the atmosphere. Warming will indeed lift more moisture from the oceans into the air. But what if the moister, warmer air increases the efficiency of rainfall, and leaves the upper atmosphere as dry, or even dryer, than it was before? We have absolutely no evidence to demonstrate that the upper atmosphere is retaining more water vapor to amplify the CO_2.

To the contrary, a team of researchers from NASA and MIT recently discovered a huge vertical heat vent in the Earth's atmosphere. It apparently increases the efficiency of rainfall when sea surface temperatures rise above 28° C. This effect seems to be big enough to vent all the heat the models predict would be generated by a doubling of CO_2.[16]

In 2001, NASA issued a press release about the heat vent discovery and the failure of the climate models to duplicate it but it attracted little media attention.[17]

THE INHERENT DANGERS OF THE KYOTO PROTOCOL

Early in my career, I served as a missionary in Africa. I lived upcountry with people who did not have access to useful energy. . . . I watched as women walked in the early morning to the forest edge, often several miles away, to chop wet green wood for fuel. . . . They became beasts of burden as they carried the wood on their backs on the return trip home. . . . Burning wood and dung inside the homes for cooking and heat created a dangerously polluted indoor atmosphere for the family. I always thought that if each home could be fitted with an electric light bulb and a microwave oven electrified by a coal-fired power plant, several good things would happen. The women would be freed to work on other, more productive, pursuits. The indoor air would be much cleaner so health would improve. Food could be prepared more safely. There would be light for reading and advancement. Information through television or radio would be received. And the forest with its beautiful ecosystem could be saved.[18]

The Kyoto Protocol will likely cost at least $150 billion a year, and possibly much more. UNICEF estimates that just $70–80 billion a year could give all Third World inhabitants access to the basics like health, education, water and sanitation.[19]

The Kyoto Protocol would probably double First World energy costs before 2012, and might quadruple them after that year. Kyoto would thus impair or even cancel out the enormous beneficial effects of technology in people's lives.

The myths of "free" wind and solar power continue to fascinate journalists and activists. Kyoto proponents assert that "renewable" energy sources will not only be adequate for the needs of modern society, but the shift from fossil fuels to solar and wind will "create jobs." This is like claiming that repairing a broken window makes us richer; instead, it just gets us back to where we had been. A shift to renewable fuels would certainly create jobs, but it would also require time and talents that could have produced additional well-being.

Energy experts note that wind and solar power are not very reliable; they are only generated when the wind is right or the sun is shining, and they are difficult to store. Backup nonrenewable power plants are needed in "spinning reserve" so the traffic lights and hospital operating theaters don't go dark. Despite decades of heavy subsidies, solar and wind power provide only about 0.5 percent of current U.S. electricity, and almost none of our transport energy.

Solar and wind are still four to ten times as expensive as fossil and nuclear energy sources. Shifting to "renewables" would also force us to convert hundreds of millions of forest and wildlands acres to windmill farms, solar panel arrays, biofuel crops, and the like.

In fact, a team of energy experts from various academic, government, and private sector research units said in a 2002 "Science Compass" article for the journal *Science*, that it would be a Herculean task to replace the fossil energy supplies any time soon.[20]

The biggest problem is that the world's current 12 trillion watt-hours of energy used per year (85 percent of it fossil-fueled) will need to be expanded to 22–42 trillion over the next fifty years in order to accommodate the world's growing population and provide economic growth for developing countries.

Energy experts say that even nuclear power will not be enough, due to a shortage of uranium ore. We will need safe nuclear breeder reactors and, ultimately, fusion power. That means developing very expensive energy technology we don't yet have.

Modern technology has also been humanity's strongest environmental conservation tool. For thousands of years, humans lived by preying on wildlife. Then we invented farming, and took most of the world's good land away from Nature for crops and pastures. Only in the past half-century has most of the world adopted the high-yield farming that permits more food to be grown per farming acre. That high-yield farming has forestalled the need to plow down millions more square miles of forests. Without the nitrogen captured from the air by fertilizer factories powered by natural gas, we'd have to clear the world's remaining forests for low-yield crops—tomorrow.

The rural half of India's population is scouring its forests daily for scarce firewood, and cutting trees it doesn't replant. India's demand for firewood is likely to double again by 2020.

Most of the Third World is already in the most polluting phase of industrial development, with grimy cities burning huge amounts of coal and smelting lots of iron for heavy machinery. The Third World needs economic growth, not another Great Depression, to move up to the cleaner industries and technologies used by the First World.

World Bank researchers have now concluded that, while the early stages of economic growth are harmful to the environment, the later stages of economic growth are environmentally constructive. Incomes above $5,000 to $8,000 per capita, where Brazil and Malaysia are today, trigger more investments in air and water cleanup and high-yield farming. They also bring massive reductions in industrial emissions, and the creation of more parks and wildlife preserves that are actually policed to protect the wildlife.[21]

Bjorn Lomborg, author of *The Skeptical Environmentalist*, recently convened a panel of leading economists to propose the most effective ways the world could use $50 billion to benefit humankind. The panel was cosponsored by the Danish government and *The Economist*. This "Copenhagen Consensus" recommended that the money be spent, first, on combatting new cases of AIDS ($27 billion); second, on reducing iron deficiency anemia in women and children through food supplements ($12 billion); third, on controlling the malaria ($13 billion) that afflicts 300 million people and kills 2.7 million annually. (The malaria control will necessarily involve the indoor use of DDT as a mosquito insecticide.) The Copenhagen panel's fourth spending priority was on agricultural research to sustainably raise crop yields and ease the competition for land between people and wildlife.

The Copenhagen Consensus ranked the Kyoto Protocol sixteenth out of seventeen proposed ways to use the money. The panelists said Kyoto's costs would outweigh its benefits—even though they assumed significant warming driven by CO_2. If they had been aware of the physical evidence endorsing the 1,500-year climate cycle, they might have ranked Kyoto even lower.

"The panel's findings are a reproach to many European leaders and to left-wing environmentalists, health activists, and anti-globalists, whose sloganeering has dominated much of the discussion of global welfare issues," wrote syndicated columnist James Glassman. "This report—sober, nonpartisan, and compassionate, with an emphasis on sound science and economic cost-benefit analysis—makes the noisy radicals look foolish."[22]

We don't necessarily subscribe to the rankings of the Copenhagen Consensus but agree that the Kyoto Protocol should be near the bottom.

WHY FEAR GLOBAL WARMING?

History, science, and our own instincts tell us that cold is more frightening than warmth. It is a psychological mystery why comfortable First World residents, armed for the first time in all of history's warmings with air conditioning, have chosen to fear "global warming."

Of course, the advocates of man-made warming have attempted to bolster a scientifically weak case with a number of essentially baseless scary scenarios.

Rising Sea Levels Will Flood Cities and Cropland, and Submerge Islands

Judging from measurements made on corals, sea levels have been rising steadily since the peak of the last Ice Age about 18,000 years ago. The total rise since then has been four hundred feet. The sea levels rose fastest during the Holocene Climate Optimum, when the major ice sheets covering Eurasia and North America melted away. For the last 5,000 years or so, the rate of rise has been about seven inches per century. Tide gauge data from the past century show a rise of about six inches—even after the strong warming between period 1920 and 1940.

When the climate warms, ocean waters expand and glaciers melt, so sea levels rise. But a warmer ocean evaporates more water, some of which ends up as snow and ice on Greenland and on the Antarctic continent, and that makes sea levels fall. More warming and more evaporation are adding ice to the Antarctic ice cap. Thus, there is no reason to expect any big acceleration of sea level increase in the twenty-first century. Researchers say it would take another 7,000 years to melt the West Antarctic Ice Sheet—a small fraction of all the ice—and we're almost sure to get another ice age before then.

More than a Million of the World's Wild Species Will Go Extinct in the Next Century

We know that species can adapt to abrupt global warming because the climate shifts in the 1,500-year cycle have often been abrupt. Moreover, the world's species have already survived at least six hundred such warmings and coolings in the past million years.

The major effect of global warming will be more biodiversity in our forests, as most trees, plants, birds, and animals extend their ranges. This is already happening. Some biologists claim that a further warming of 0.8 degrees Celsius will destroy thousands of species. However, the Earth warmed much more than that during the Holocene Climate Optimum, which occurred 8,000 to 5,000 years ago, and *no known species were driven extinct* by the temperature increase.

There Will Be More Hunger and Famine as Fields Become Too Hot to Grow Crops

High-tech farming, not climate, has governed the world's overall food production since the seventeenth century. There will be little temperature change in the tropics, where food production is still inadequate. The northern plains in Canada and in Russia will become warmer, and will produce more food. Modern society can help make tropical farming high-tech, or transport more food from Siberia to people with new nonfarm jobs in India or Nigeria. Any famines will be humanity's fault, not the fault of the climate.

There Will Be More Storms and Worse Storms with Global Warming

There has been no increase in the frequency or severity of hurricanes, blizzards, cyclones, tornadoes, or any other kind of storms during the warming of the past 150 years. That makes sense, because storms are driven by the temperature differential between the equator and the polar regions. Since greenhouse warming should boost the temperatures at the poles much more than at the equator, warming will reduce the differential and moderate the storms. History and paleontology tell us the warmings have experienced better, more stable weather than the coolings.

Global Warming Will Trigger Abrupt Global Cooling

The warming activists claim that increased meltwater due to higher temperatures could overwhelm the Great Atlantic Conveyor, the huge ocean current that distributes heat from the equator to the poles. The Gulf Stream would then shut down, and we would all be covered in ice before you can say "carbon dioxide." It happened once before—but then the world had trillions of additional tons of ice in Canadian and Siberian ice sheets for the warming to melt. The climate models—surprise!—tell us that it won't happen during the Modern Warming, because the Earth doesn't have enough ice left.

Human Death Tolls Will Rise with the Heat, Insects, and Disease of Global Warming

Freezing weather kills far more people than hot weather, and there'll be less freezing winter weather during the Modern Warming. As for mosquito-borne diseases, window screens and insecticides wiped out most of the world's malaria and yellow fever, not cold weather. The world's biggest malaria outbreak was in Russia in the 1920s.

Coral Reefs Will Die Out with Warming

Many coral reefs have "bleached" (lost the algae that live in symbiotic partnership with them) when sea temperatures rose. But they also bleach when temperatures fall. That's because the corals partner with the algae varieties best adapted to their current temperature. When the water warms, they eject their cold-water partners and welcome warmer-weather friends, and vice versa. That's how they have survived for millions of climate-varied years.

FEAR THE NEXT ICE AGE

The climate event that deserves real concern is the next Big Ice Age. That is inevitably approaching, though it may still be thousands of years away. When it comes, temperatures may plummet 15 degrees Celsius, with the high latitudes getting up to 40 degrees colder. Humanity and food production will be forced closer to the equator, as huge ice sheets expand in Canada, Scandinavia, Russia, and Argentina. Even Ohio and Indiana may gradually be encased in mile-thick ice, while California and the Great Plains could suffer century-long drought.

Keeping warm will become the critical issue, both night and day. Getting enough food for eight or nine billion people from the relatively small amount of arable land left unfrozen will be a potentially desperate effort. The broad, fertile plains of Alberta and the Ukraine will become sub-Arctic wastes. Wildlife species will be extremely challenged, even though they've survived such cold before—because this time there will be more humans competing for the ice-free land.

That's when human knowledge and high-tech farming will be truly needed.

In contrast, none of the scary scenarios posited by today's global warming advocates took place during the Earth's past warm periods.

Why have humans chosen to panic about the planet returning to what is very probably the finest climate the planet has known in all its millions of years? Is it simply guilt because climate alarmists told us we humans were causing the change?

If so, then it becomes all the more important to check their evidence.

NOTES

1. W. Dansgaard et al., "North Atlantic Climatic Oscillations Revealed by Deep Greenland Ice Cores," in *Climate Processes and Climate Sensitivity*, ed. F. E. Hansen and T. Takahashi (Washington, D.C.: American Geophysical Union, 1984), Geophysical Monograph 29, 288–98.

2. John C. Ryan, "Greenhouse Gases on the Rise in the Northwest," Northwest Environment Watch, 1995, <www.northwestwatch.org> (12 February 2004).

3. Art Bell and Whitley Strieber, *The Coming Global Superstorm* (New York: Pocket Books, 2000), 10–11.

4. Former U.S. Senate Majority Leader George Mitchell (D-Maine), *World on Fire: Saving an Endangered Earth* (New York: Scribner, 1991), 70–71.

5. Natural Resources Defense Council, www.nrdc.org/globalWarming/ (27 May 2003).

6. Sir John Houghton, Chairman, Scientific Assessment Working Group, United Nations Intergovernmental Panel on Climate Change, letter to the World Council of Churches, 1996.

7. Christine Stewart, then Canadian Minister of the Environment, before the editors and reporters of the *Calgary Herald*, 1998, and quoted by Terence Corcoran, "Global Warming: The Real Agenda," *Financial Post*, 26 December 1998, from the *Calgary Herald*, 14 December 1998.

8. Paul Recer, "Study of Tree Rings Shows Earth Has Normal Cycles of Warmth, Cooling," Associated Press, 22 March 2002.

9. John Christy, Professor of Atmospheric Science, University of Alabama–Huntsville and a lead author for the UN Intergovernmental Panel on Climate Change, speaking before the House of Representatives committee, 13 May 2003.

10. William H. Hooke, policy director, American Meteorological Society, "Avoiding a Catastrophe of Human Error," *Washington Post*, 5 January 2005.

11. Thomas Gale Moore, "Why Global Warming Would Be Good for You," *The Public Interest* (Winter 1995): 83–99.

12. Gerard Bond et al., "Persistent Solar Influence on North Atlantic Climate during the Holocene," *Science* 294 (16 November 2001): 2130–136.

13. Richard Kerr, "A Variable Sun Paces Millennial Climate," *Science* 294 (16 November 2001): 1431–433.

14. Frederick Seitz, former president, National Academy of Sciences, "A Major Deception on Global Warming," *Wall Street Journal*, 12 June 1996, editorial page. S. Fred Singer, *Climate Policy from Rio to Kyoto: A Political Issue for 2000 and Beyond* (Palo Alto, CA: Hoover Institution, Stanford University, 2000), 19.

15. D. H. Douglass, B. Pearson, and S. F. Singer, "Disparity of Tropospheric and Surface Temperature Trends: New Evidence," *Geophysical Research Letters* 31: L13207.doi:10.1029/2004/GL020212 (2004).

16. Richard Lindzen, Ming-Dah Chou, and Arthur Hou, "Does the Earth Have an Adaptive Infrared Iris?" *Bulletin of the American Meteorological Society* 82 (2001): 417–32.

17. "Natural 'Heat Vent' in Pacific Cloud Cover Could Diminish Greenhouse Warming," press release, NASA Goddard Space Flight Center, 28 February 2001.

18. John Christy, climatologist, University of Alabama–Huntsville.

19. Bjorn Lomborg, *The Skeptical Environmentalist* (London: Cambridge University Press, 2001), 322.

20. M. I. Hoffert et al., "Science Compass: Advanced Technology Paths to Global Climate Stability: Energy for a Greenhouse Planet," *Science* 298 (2002): 981–87.

21. *International Trade and the Environment*, World Bank Discussion Paper 159, Patrick Low, ed. (January 1992); also, G. Grossman and A. Kreuger, "Economic Growth and the Environment," *The Quarterly Journal of Economics* 370 (1995): 353–77.

22. James Glassman, "How to Save the World," *Washington Times*, 9 June 2004, A16.

Chapter Two

How Did We Find the Earth's 1,500-year Climate Cycle?

Global Warming Advocates Say:

"Removal of all forcing except greenhouse gases from the 1000-year time series results in a residual with a very large late 20th century warming that closely agrees with the response predicted from greenhouse gas forcing. The combination of a unique level of temperature increase in the late 20th century and improved constraints on the role of natural variability provides further evidence that the greenhouse effect has already established itself above the level of natural variability in the climate system."[1]

Reality-Based Skeptics Say:

"George Washington's winter at Valley Forge in 1777–1778, when temperatures fell as low as –15°C, was relatively mild for those days; some years, New York Harbor froze solid. Indeed, so bitter were the centuries from about 1400 until 1900 that they have been dubbed 'The Little Ice Age.' But new evidence appears to confirm that the long cold snap was nothing exceptional. Instead, it was only the most recent swing in a climate oscillation that has been alternately warming and cooling the North Atlantic region, if not the globe, for ages upon ages."[2]

"Although first seen in the Greenland ice cores, the Dansgaard-Oeschger [1,500-year cycle] events are not [just] a local feature of Greenland climate. . . . [S]ubtropical sea surface temperatures in the Atlantic closely mirror the sequence of events in Greenland. Similar records have been found near Santa Barbara, California, in the Cariaco Basin of Venezuela, and off the coast of India."[3]

In 1983, Denmark's Willi Dansgaard and Switzerland's Hans Oeschger were among the first people in the world to see two ice cores nearly a mile long that were then being brought up from deep holes in the Greenland Ice Sheet—bringing up with them 250,000 years of the Earth's layered climate history. Dansgaard and Oeschger then began the frigid, laborious task of counting the layers and analyzing their contents to reveal Earth's past climatic changes.

Over the previous dozen years, the two researchers had pioneered ways to sink the sort of hollow drill core that creates oil wells—and bringing up ice cores instead of petroleum. They learned, among other things, that the ratio of oxygen-18 isotopes to oxygen-16 isotopes in ice could reveal the air temperature when snowflakes fell to earth. That meant the new, long ice cores (one from the north end of the big island, one from the south) were a record of Greenland's temperatures almost year by year for over 2,500 centuries. No such long-term climate history had ever before been available.

They expected to see the big ice ages in the ice core's record. When it came up, however, Dansgaard and Oeschger were startled to find, superimposed on the big ice age/interglacial climate swings, a smaller, moderate, and more persistent temperature cycle. They estimated that the average cycle length was 2,550 years. The timing of the cycles seemed to match closely with the known history of recent glacier advances and retreats in northern Europe.

THE DISCOVERY

The report that Dansgaard and Oeschger wrote in 1984, "North Atlantic Climatic Oscillations Revealed by Deep Greenland Ice Cores," was almost eerie in its accuracy, its completeness, and its logical linkage of the moderate climate cycles to the sun.[4] The only major correction imposed by subsequent research has been an adjustment for even more cycles than Dansgaard and Oeschger first counted. The average length of the cycles has now been shortened by almost half—from their original estimate of 2,550 years to 1,500 years (plus or minus 500).

Dansgaard and Oeschger were correct when they told us that the climate shifts were moderate, rising and falling over a range of about 4 degrees Celsius in northern Greenland, but only half a degree Celsius when averaged over the Northern Hemisphere. They noted that the cycles were confirmed by:

1. Their appearance in two different ice cores drilled more than 1,000 miles apart;

2. Their correlation with known glacier advances and retreats in northern Europe data from a seabed sediment core that had been drilled in the floor of the Atlantic Ocean west of Ireland a decade earlier.[5]

They sought an explanation. They suggested that the cycles were not as readily observed in the Southern Hemisphere due to the major differences in the land–sea distribution. They dismissed volcanoes as a causal factor because there is no such cycle in volcanic activity. They noted that the cycle shifts were abrupt, sometimes gaining half of their eventual temperature

---

CYCLES WITHIN CYCLES WITHIN CYCLES

The ice ages, which have dominated 90 percent of the Earth's climate for the past several million years, are thought to have been the result of the Earth's eccentric orbit around the sun. The shape of the Earth's orbit becomes more elliptic and then less, on a cycle of about 100,000 years. This alters the distance from the Earth to the sun, forcing the sun's radiation to travel 3 percent farther to reach our planet during the ice ages and reducing its intensity. The Earth's orbit is currently not very elliptical, and the variation between January's solar radiation and July's is only about 6 percent. When the orbit is more elliptical, the energy received can vary as much as 20 to 30 percent.

The Earth also has an axial tilt, which is why the map globe on my desk tilts at an angle of about 23 degrees instead of sitting straight up with the North Pole at the top. The angle at which the sun strikes the earth's surface varies with latitude and season—and with the degree of tilt that varies on its own cycle of about 41,000 years. At present, the axial tilt is about in the middle of its range.

Finally, the Earth does a slow wobble as it spins on its axis. This wobble occurs on a cycle of 23,000 years. This "precession" is significant to the climate when the North Pole is pointed toward the star Vega, one of the brightest and nearest "landmarks" to Earth in the solar system. At that point, the Earth gets both harsh winters and hot summers. At the other extreme of the wobble when the precession puts summer at the "perihelion," the Earth gets milder summers and more winter ice sheets survive to the next winter. We now should be entering an extended period of ice sheet growth, according to the wobble cycle.

If all of this seems complicated, it is. (No wonder the simplistic Greenhouse Theory caught on with the public.) Climate forecasters must factor the 100,000-year elliptical cycle, the 41,000-year axial tilt cycle, and the 23,000-year precession or "wobble" cycle, plus the 1,500-year solar-driven cycle.[6] However, it is the 1,500-year cycle that drives most of the Earth's climatic change during interglacial periods like this one.

change in only a decade or so; that suggested an external forcing, perhaps amplified and transmitted globally by the ocean currents and winds. They wrote:

> Since the solar radiation is the only important input of energy to the climatic sys-
> tem, it is most obvious to seek an explanation in solar processes. Unfortunately
> we know much less about the solar radiation output than about the emission of
> solar particulate matter in the past.[7]

The two scientists did know, however, that both carbon-14 and beryllium-10
isotopes vary inversely with the strength of solar activity. The isotopes of
both elements in their Greenland ice cores showed historic lows during the
Maunder Solar Minimum (1645–1715)—the absolute coldest point of the Little Ice Age.[8]

Their speculation was also informed by known warming and cooling periods: the Roman Empire was known to have flourished in a relatively warm
climatic period and collapsed during the colder Dark Ages; the Medieval
Warming (950 to1300), a period of abundant crops and stable weather, and
the Little Ice Age itself, a relatively cold period from 1300 to 1850, are widely
recorded in human histories. They concluded that "the abrupt temperature rise
in the 1920s may thus be the latest member of a very long series of similar
events that occurs once every ca. 2550 years to an extent that is modulated by
the degree of glaciation and dependent on the latitude."[9]

OTHERS FIND THE 1,500-YEAR CYCLE

The importance of the 1,500-year cycles increased dramatically four years
later when such evidence was also found at the other end of the world—in an
ice core from the Antarctic's Vostok Glacier. The discovery was made by a
French and Russian research team, and the French team's leader, Claude Lorius, eventually shared the Tyler Prize with Dansgaard and Oeschger.[10]

The scientific world had known about the sunspot connection to Earth's
climate for some four hundred years, though we had never understood how it
worked. The connection between the depths of the Little Ice Age and the
Maunder Solar Minimum was obvious even to seventeenth-century observers
with nothing but crude telescopes.

In 1991, E. Friis-Christensen and K. Lassen noted that the correlation between solar activity and northern hemisphere land temperatures was even
stronger if one used the length of the solar cycle to represent the sun's variations instead of the number of sunspots.[11] Their paper in *Science* was titled,
"Length of Solar Cycle: An Indicator of Solar Activity Closely Associated

with Climate." They concluded that the solar connection explained 75 to 85 percent of recent climate variation.

In 1996, Lloyd Keigwin reported finding the 1,500-year cycle in the sea surface temperatures of the Sargasso Sea, reconstructed from the oxygen isotopes in the tiny one-cell organisms of a seabed sediment core.[12]

During the past decade, numerous researchers have found the cycle in many long-term temperature proxies, particularly from isotopes of oxygen, carbon, beryllium, and argon trapped in glacier ice, from fossil pollen records, and from algae cyst assemblages in lake and seabed sediments.

At first, some researchers contended that the cycle might be driven by changes in the winds and ocean currents that distribute the sun's heat around the globe. However, they couldn't find any internal cause for the sudden, long-term shifts from warm to cold and back again.

The researcher who has done the most to make the public aware of the 1,500-year climate cycle is Gerard Bond of Columbia University's Lamont-Doherty Earth Observatory. He did it by analyzing ice-rafted debris from sediments on the floor of the southern area of the North Atlantic Ocean. He found that, roughly every 1,500 years, there was a surge in the amount of rocky bits picked up by the glaciers as they ground their way across eastern Canada and Greenland, and this ice-rafted debris was floated much farther south before it dropped to the sea floor as the icebergs melted. Both the increase in the volume of debris and its floating much farther south indicate severe cold periods.

Bond then did a follow-up study, counting ratios of carbon and beryllium isotopes in the sediments and found these solar proxies correlated very closely with the cycles in the iceberg debris. He found nine of these cycles in the last 12,000 years.

Bond's 1997 research report in *Science* begins: "Evidence from North Atlantic deep-sea cores reveals that abrupt shifts punctuated what is conventionally thought to have been a relatively stable Holocene [interglacial] climate. During each of these episodes, cool, ice-bearing waters from north of Iceland were advected as far south as the latitude of Britain. At about the same times, the atmospheric circulation above Greenland changed abruptly. . . . Together, they make up a series of climatic shifts with a cyclicity close to 1470 years (plus or minus 500 years). The Holocene events, therefore, appear to be the most recent manifestation of a pervasive millennial-scale climatic cycle *operating independently of the glacial-interglacial climate state*"[13] (emphasis added).

Bond's team analyzed two deep-seabed cores, from opposite sides of the North Atlantic, which took them back 30,000 years into prehistory, into the Ice Age itself. They used high-resolution mass spectrometers to carbon-date the plankton fossils (and thus the layers) in the sediments.

The proxies definitely indicated a series of ice intrusions big enough to deliver increased iceberg sediments to two southerly sites more than 1,000 kilometers apart.

Bond's cycles matched those in the cores from the Greenland Ice Sheet, very much strengthening our confidence that the cycles are real and significant. Bond's subsequent study, published in *Science,* demonstrated the linkage between the Earth's warming–cooling cycle and the sun, using carbon-14 and beryllium-10 as proxies for solar warming/cooling.[14]

Ulrich Neff of the Heidelberg Academy of Sciences found Bond's Atlantic seabed cycles replicated in a cave stalagmite on the distant Arabian Peninsula.[15]

According to L. Keigwin,

The most recent drift-ice cycle, between about 100 and 1,100 years ago, corresponds to temperature cycles in the Sargasso Sea and in the upwelling region off West Africa suggesting a broad regional response in the subpolar and subtropical North Atlantic. The interval also encompasses a series of marked increases in surface ocean upwelling in the Cariaco Basin [off the coast of Venezuela], prominent changes in lake levels in equatorial east Africa, and distinct episodes of drought in the Yucatan Peninsula, all of which have been linked to episodes of reduced solar irradiance. . . . The solar-climate links implied by our record are so dominant over the last 12,000 years that it seems almost certain that the well-documented connection between the Maunder Solar Minimum and cold decades of the Little Ice Age could not have been a coincidence."[16]

A CLIMATE CYCLE PROXY
FROM OTHER PARTS OF THE WORLD

Keigwin's reference to a study in "the upwelling region off West Africa" takes us to the work of Bond's Lamont-Doherty colleague, Peter deMenocal. deMenocal led a team studying plankton fossils and airborne dust in a deep-sea core from the Atlantic coast of Africa, at Cap Blanc, Mauritania.[17] Cap Blanc is at the normal boundary between the colder subpolar waters of the north Atlantic and the warmer tropical water masses of the southern Atlantic. It also accumulates sediment rapidly (important for dating accuracy) because it has a high rate of plankton productivity due to the nearby upwelling, and because it lies within the plume of dry-season mineral dust blown from Africa.

deMenocal's results confirmed the same set of cycles identified by Bond's iceberg debris in the North Atlantic, this time thousands of kilometers closer to the equator, where ice never forms.

The deMenocal team's deep-sea cores document a history of major changes in sea surface temperatures off West Africa, which are linked to the same pattern of climate change Bond found in the North Atlantic—even though land temperatures near the equator don't change much during the cycle. As we shall find, the amount of rainfall changes with the cycle, instead of the thermometer readings.

Changes in plankton numbers and species gave the deMenocal team ocean temperature readings from the past, and the amounts of dust blown from Africa were an indicator of drought. The proxies tell us that when the sea surface temperatures fell off West Africa, much of the continent went drier for centuries. Then the climate snapped back again, quickly, bringing such heavy rains that many large lakes formed in the Sahara Desert. The most recent cooling in the region was a two-stage Little Ice Age between 1300 and 1850, essentially simultaneously with similar coolings in the Greenland ice cores, in the seabed sediments of the North Atlantic found by Bond, and in the reconstructed sea-surface temperatures of the Sargasso Sea.[18]

During the Medieval Warming, deMenocal's team found dryness, and then a wet period during the two-stage Little Ice Age.

Dirk Verschuren, of Belgium's Ghent University, worked with a 1,100-year sediment core from an East African lake. His team also found a drier African climate during the Medieval Warming, and a wetter Little Ice Age—but the cold, wet period was interrupted by three prolonged droughts.

Bond concluded that every 1,500 years, harsh cold periods drop North Atlantic ocean temperatures by 2 to 3.5 degrees Celsius. deMenocal says that ocean temperatures off Africa simultaneously dropped even more sharply than in the North Atlantic, with changes of 3 to 4 degrees Celsius.

deMenocal and Bond show us a dynamic climate system in which temperature and rainfall constantly change. Moreover, the cycle they have found long predates human industrial activity, and is linked to the variability of the sun's activity. The two proxy studies strongly confirm each other and accord with the ice cores. This virtually destroys the old idea that our climate changes slowly and predictably between Ice Ages.

The pollen fossils in the North American Pollen Data Base records show a major reorganization of the vegetation across North American nine times in the past 14,000 years "with a periodicity of 1650 years plus or minus 500 years."[19] How close is that to an erratic 1,500-year cycle?

Analysis was carried out by a team led by Andre E. Viau of the University of Ottawa. They wrote: "We suggest that North Atlantic millennial-scale climate variability is associated with rearrangements of the atmospheric circulation with far-reaching influences on the climate."[20]

Sediments from a lake in southwestern Alaska reveal the same cyclic variations in climate and ecosystems that Bond found in the North Atlantic and deMenocal found off the coast of West Africa. The University of Illinois' F. S. Hu led a team that analyzed the silica produced by living organisms, organic carbon and organic nitrogen to reconstruct the temperature history.

Their findings are evidence that the climate shifts have been similar in the subpolar regions of both the North Atlantic and North Pacific—"possibly because of sun-ocean-climate linkages."[21]

In the Sulu Sea near the Philippines, the productivity of the phytoplankton is closely related to the strength of the winter monsoon. The production of phytoplankton was larger during glacial periods than during interglacial periods, but the researchers found that "the 1,500-year cycle . . . seems to be a pervasive feature of the monsoon climatic system."[22]

Andersson and his team constructed a 3,000-year temperature history for the Norwegian Sea from the stable isotopes in the plankton and the number and types of protozoan skeletons from seabed sediment cores.[23] The climate history shows a long cold period before the Roman Warming, then the Dark Ages, the Roman Warming, and the Little Ice Age. Andersson also notes that "surface ocean conditions warmer than present were common during the past 3,000 years."[24]

Three stalagmites in a cave in Sauerland, Germany, give a climate history for more than 17,000 years, clearly tracing backward through the Little Ice Age, the Medieval Warming, the Dark Ages cold period, the Roman Warming—and the unnamed cold period that preceded the Roman Warming.[25]

Stefan Niggemann of the Heidelberg Academy of Science and his coauthors specifically note that their stalagmite temperature records "resemble records from an Irish stalagmite."[26] McDermott's stalagmite, in turn, confirms the Sauerland stalagmites.[27]

In a chronology of landslides in the Swiss Alps, the three most recent and best-documented periods of landslides (colder and wetter weather) were during the unnamed cold period before the Roman Warming (600–200 B.C.), the Dark Ages (A.D. 300–850), and the Little Ice Age (1300–1850).[28]

In the Arabian Sea, west of Karachi, Pakistan, two seabed sediment cores date back nearly 5,000 years, and show "the 1470-year cycle previously reported from the glacial-age Greenland ice record."[29] The authors, W. H. Berger and U. von Rad, suggest the cycles are tide-driven, because so many of them were multiples of basic tidal cycles. However, they also note that *"internal oscillations of the climate system cannot produce them"* (emphasis added). They see the Modern Warm Period as another in a series of externally forced cycles.[30]

THE EARTH'S CLIMATE: CYCLING A MILLION YEARS AGO AND BEYOND

Maureen Raymo[31] of Boston College says that the Earth was undergoing Dansgaard-Oeschger climate cycles more than a million years ago. Raymo and her research team retrieved a very long sediment core from the deep-sea bottom south of Iceland. The sediments show the same pattern of periodic surges in ice-rafted debris found by Gerard Bond, but coming from a period much farther back in time.

So our newly discovered 1,500-year cycle goes back a million years. It functioned through Ice Ages. Why should it stop in our era?

BUT THERE IS NO 1,500-YEAR SOLAR CYCLE

"Researchers from several institutes in Germany . . . used a computer model to show that small changes in the Sun could have triggered a series of abrupt warmings in the last Ice Age. . . . Some researchers have speculated that variations in the Sun could have triggered the 1,470-year cycle. The [sunspot observations] show solar cycles with periods of about 87 and 210 years. A cycle of 1,470 years, however, has not been found so far. . . . Because they are close to factors of 1,470 years, the 210-year cycle and the 87-year cycle of the Sun could combine to form a period of 1,470 years and thus explain the climate cycle of the ice-age."[32]

The Earth, and especially the North Atlantic region, showed a 1,470-year climate cycle. The sun did not. The sun clearly had the 87-year Gleissberg cycle and the 210-year DeVries-Suess cycle, but no 1,470-year cycle had been found.

Holger Braun's colleagues at the Potsdam Institute for Climate Impact Research had already published an article on a possible cycle mechanism, speculating that changes in ocean circulation during the ice age could explain the abrupt climate shifts.[33] However, no one had explained the regularity of the phenomenon.

During the last ice age, the Greenland ice showed that the cycle varied less than 10 percent in its timing—through at least twenty events over tens of thousands of years. That implies a strong external forcing, and brings us back to Dansgaard and Oeschger's original speculation in 1984 that the sun was the forcing agent. And, don't forget the statement by Gerard Bond that "over the last 12,000 years virtually every centennial time-scale increase in drift ice documented in our North Atlantic records was tied to a solar minimum."[34]

But how?

During the current Holocene Warming, the cycle has clearly been less regular: the Holocene Climate Optimum lasted about 4,000 years—from 9,000 to 5,000 years ago. The Dark Ages (600–950) and the Medieval Warming that followed it (950–1300) lasted a total of only 700 years combined. Still, the cycle was clearly operating through the Roman Warming, the Dark Ages, the Medieval Optimum, the Little Ice Age—and, dare we say, the Modern Warming?

The cycle is obviously continuing to function. It could be less immediately powerful in its impacts if the Earth doesn't have big ice sheets with which to generate big surges of fresh meltwater quickly. But it's still there.

Holger Braun and his colleagues speculated that the known 87- and 210-year solar cycles must be superimposed on each other. Both numbers are close to prime factors of 1,470. Seven of the longer 210-year solar cycles and seventeen of the shorter 87-year solar cycles fit neatly into the Earth's 1,470-year climate cycle. If those two independent cycles were occurring simultaneously, they might either add to or cancel each other out, creating the longer, and more complex 1,470-year climate shift.[35]

The team ran their idea on a computer model. It worked. They report that "an intermediate-complexity climate model with glacial climate conditions" simulated the rapid 1,470-year climate shifts of the Dansgaard-Oeschger events "when forced by periodic freshwater input into the North Atlantic ocean in cycles of 86 and 210 years."[36]

The researchers did not try to present the mechanisms of the solar forcing. They lacked key information: a reliable and detailed reconstruction of solar activity over the tens of thousands of ice age years; the details of the atmospheric chemistry; and the detailed dynamics of continental ice sheets and sea ice. They demonstrated, however, that "the glacial 1,470-year climate cycles could have been triggered by solar forcing despite the absence of a 1,470-year solar cycle."[37]

Their paper says that the 1,470-year cycles ended with the Ice Age, but the Earth's moderate, natural warmings and coolings have clearly continued. One of the Potsdam authors, Stefan Rahmstorf, has elsewhere noted that "the so-called 'Little Ice Age' of the 16th–18th centuries may be the most recent cold phase of this cycle."[38]

Is this the final endorsement of the 1,500-year climate cycle as the driving force in our Modern Warming?

ASSESSING THE PREHISTORIC
EVIDENCE OF CLIMATE CHANGE

Now let's assess what we know of this mysterious recurring climate event. *First*, we know that it's big. It can be traced from the North Atlantic to the equator off Africa, and from Alaska to the Philippines and Antarctica. That's no small set of tracks.

Second, we know that the events stem from natural causes. The footprints go back more than a million years, long before any human activities affected the climate.

Third, we know that the climate forcing is powerful enough to drive itself right through the ice ages. When your driving force can warm the earth despite trillions of tons of extra ice covering the northern hemisphere continents, it is powerful indeed. It must, in fact, be one of the most powerful forces humanity has ever comprehended.

Fourth, we know from the isotopes and history that the cycle is moderate. We know that our ancestors and the ancestors of our wild species (most of them millions of years old) have survived a whole series of these climate cycles—with neither central heating nor air conditioning. Primitive people survived even the harsh ice ages by simply moving out of the way of rising waters and encroaching glaciers. Apparently, much of the wildlife did the same, and those species that couldn't move reseeded themselves when the ice and/or the stifling heat left. We know that the massive species extinctions in the Earth's past have not been linked to internal temperature changes but to such external events as asteroid collisions and comet near-misses. They blotted out the sun for years at a time with atmospheric debris.

The typical climate shift produced by the cyclical forcing has a beginning, often involving a sudden and erratic climate shift of up to 2 degrees Celsius at northern mid-latitudes, and with temperatures shifting even more in the Arctic. It has a middle that is climatically stable in the warm periods and unstable during the "ice ages." Then the cycle phase has an end. When the Medieval Warming ended abruptly, the huge storm surges that hit Europe at the beginning of the Little Ice Age should definitely worry us. The problem will be to forecast when the current warming phase will end—at a date now wholly uncertain. That's when we'll need to decide which investments to make in storm surge protection, and where.

The key thing for us all to remember is that the 1,500-year climate cycle is not an unproven theory like the model-based predictions for the Greenhouse Theory. The 1,500-year climate cycle is *real,* based on a wide variety of physical evidence from around the globe.

The ice cores were cut from real-world ice sheets built up into layers over thousands of years. The satellites actually measured the sun's varying rays. The mass spectrometers actually counted the isotopes from the cores that confirmed the pattern of solar variation. The sunspot counts of the last four hundred years are handwritten on the yellowed pages of the observers' diaries. The Armagh Observatory's solar record has been carefully kept daily for more than two hundred years. The flares on the sun are recorded on film.

The tree rings are there to be counted and recounted. The sediment cores are in storage, awaiting further research. The heavy-oxygen isotopes are demonstrably different from the lighter ones. The midges whose heads are found in the sediments actually lived. The pollen grains fell from plants, recently or long ago, but the plants were alive. The stalagmites patiently built up over thousands of years.

There's no 1,470-year solar cycle. However, the Holger Braun computer model run found that the sun's well-known 87-year and 210-year cycles, when superimposed, could create the longer 1,470-year cycle.

None of this climate cycle evidence is as likely to mislead as the unverified computer models that have received so much funding and media attention during the "greenhouse years."

Dansgaard, Lassen, and Bond all argue that the force behind the cycles is solar. Berger and von Rad argue that "internal oscillations of the climate system cannot produce" the quick-changing 1,500-year cycles. Jan Veizer and Nir Shaviv agree that the forcing producing the 1,500-year cycle is extraterrestrial, but add in the Milky Way and other galactic sources of cosmic rays.

The more we learn about the 1,500-year cycle, the less likely it seems that the recent warming is man-made—or dangerous.

NOTES

1. Thomas J. Crowley, "Causes of Climate Change over the Past 1000 Years," *Science* 289 (2000): 270–77.

2. Richard Kerr, "The Little Ice Age: Only the Latest Big Chill,'" *Science* (25 June 1999): 2069.

3. Stefan Rahmstorf, Potsdam Institute for Climate Impact Research, "Climate, Abrupt Change," in *Encyclopedia of Ocean Science*, ed. J. Steele (London: Academic Press, 2001), 1–6.

4. W. Dansgaard et al., "North Atlantic Climatic Oscillations Revealed by Deep Greenland Ice Cores," in *Climate Processes and Climate Sensitivity*, ed. F. E. Hansen and T. Takahashi (Washington, D.C.: American Geophysical Union, 1984), Geophysical Monograph 29, 288–98.

5. N. G. Pisias et al., "Spectral Analysis of Late Pleistocene-Holocene Sediments," *Quaternary Research* (March 1973): 3–9.

6. John D. Imbrie and Katherine Palmer Imbrie, *Ice Ages: Solving the Mystery* (Cambridge, MA: Harvard University Press, 1986).

7. Dansgaard et al., "North Atlantic Climatic Oscillations Revealed by Deep Greenland Ice Cores," 288–98. See also Hans Oeschger, "Long-Term Climate Stability: Environmental System Studies," *The Ocean in Human Affairs*, ed. S. Fred Singer (New York: Paragon House, 1990).

8. For a fuller discussion of the Maunder Solar Minimum, see Willie H. Soon and Steven H. Yaskell, *The Maunder Minimum and the Variable Sun-Earth Connection* (Singapore: World Scientific Publishing, 2004).

9. Dansgaard et al., "North Atlantic Climatic Oscillations Revealed by Deep Greenland Ice Cores," 288–98.

10. C. Lorius et al., "A 150,000-Year Climatic Record from Antarctic Ice," *Nature* 316 (1985): 591–96.

11. E. Friis-Christensen and K. Lassen, "Length of the Solar Cycle: An Indicator of Solar Activity Closely Associated with Climate," *Science* 254 (1999): 698–700.

12. L. Keigwin, "The Little Ice Age and Medieval Warm Period in the Sargasso Sea," *Science* 274 (1996): 1503–508.

13. G. Bond et al., "A Pervasive Millennial Scale Cycle in North Atlantic Holocene and Glacial Climates," *Science* 278 (1997): 1257–266.

14. G. Bond, "Persistent Solar Influence on North Atlantic Climate during the Holocene," *Science* 294 (2001): 2130–136.

15. U. Neff et al., "Strong Coherence between Solar Availability and the Monsoon in Oman between 9 and 6 kyr Ago," *Nature* 411 (2001): 290–93.

16. L. Keigwin, "The Little Ice Age and Medieval Warm Period in the Sargasso Sea," *Science* 274 (1996): 1504.

17. P. deMenocal et al., "Coherent High- and Low-Latitude Climate Variability during the Holocene Warm Period," *Science* 288 (2000): 2198.

18. L. Keigwin, "The Little Ice Age and the Medieval Warm Period in the Sargasso Sea," 1503–508.

19. A. E. Viau et al., "Widespread Evidence of 1,500-yr Climate Variability in North America during the Past 14,000 Years," *Geology* 30 (2002): 455–58.

20. Ibid.

21. F. S. Hu et al., "Cyclic Variation and Solar Forcing of Holocene Climate in the Alaskan Subarctic," *Science* 301 (2003): 1890–893.

22. T. De Garidel-Thoron and L. Beaufort, "High-Frequency Dynamics of the Monsoon in the Sulu Sea during the Last 200,000 Years," paper presented at the EGS General Assembly, Nice, France, April 2000.

23. C. Andersson et al., "Late Holocene Surface Ocean Conditions of the Norwegian Sea (Voring Plateau)," *Paleoceanography* 18 (2003): 10.1029/2001PA000654.

24. Ibid.

25. S. Niggemann et al., "A Paleoclimate Record of the Last 17,600 Years in Stalagmites from the B7 Cave, Sauerland, Germany," *Quaternary Science Reviews* 22 (2003): 555–67.

26. McDermott et al., 1999; Niggemann et al., 2003.

27. F. McDermott et al., "Centennial-Scale Holocene Climate Variability Revealed by a High-Resolution Speleothem O-18 Record from SW Ireland," *Science* 294 (2001): 1328–333.

28. F. Dapples et al., "New record of Holocene Landslide Activity in the Western and Eastern Swiss Alps: Implication of Climate and Vegetation Changes," *Ecologae Geologicae Helvetiae* 96 (2003): 1–9.

29. W. H. Berger and U. von Rad, "Decadal to Millennial Cyclicity in Varves and Turbidites from the Arabian Sea: Hypothesis of Tidal Origin," *Global and Planetary Change* 34 (2002): 313–25.

30. Ibid.

31. Maureen Raymo received a National Young Investigator Award from the National Science Foundation and the Cody Award from the Scripps Institution of Oceanography.

32. Potsdam Institute for Climate Impact Research press release announcing publication of study by Holger Braun et al.

33. A. Ganopolski and S. Rahmstorf, "Rapid Changes of Glacial Climate Simulated in a Coupled Climate Model," *Nature* 409 (6817) (2001): 153–58.

34. G. Bond et al., "Persistent Solar Influence on North Atlantic Climate during Holocene," *Science* 294 (2001): 2130–136.

35. H. Braun et al., "Possible Solar Origin of the 1,470-Glacial Climate Cycle Demonstrated in a Coupled Model," *Nature* 438 (2005)Z: 208–11.

36. Ibid.

37. Ibid.

38. S. Rahmstorf, "Timing of Abrupt Climate Change: A Precise Clock:" *Geophysical Research Letters* 30 (2003): 10.1029/2003GLo17115.

Chapter Three

Shattered Glass in the Greenhouse Theory

Man-Made Warming Activists Say:

"Although human-caused global warming is among the most pervasive threats to the web of life, its root cause can be addressed. The burning of fossil fuels—coal, oil, and gas—releases carbon dioxide (CO_2) into the atmosphere. This carbon pollution blankets the earth, trapping heat, and causing global warming. Reducing these emissions is the first step in stopping global warming. . . . WWF [World Wildlife Fund] is a key advocate for major reductions in CO_2 emissions. . . . A slight rise in global temperatures threatens wild animals like polar bears that rely on disappearing sea ice to access their food supply. Help save polar bears and all life on Earth by taking steps to cut emissions of CO_2. Along with WWF, you can be part of the solution."[1]

"What You Can Do to Decrease Global Warming. . . . [S]ome of the actions which we all have to take will slightly decrease your present standard of living. First, since the largest portion of electricity in the [United States] is produced by burning coal, we should try to cut down on our demand for electricity. Coal combustion creates the largest amount of CO_2 per energy unit. . . . For instance, plant several trees on the south side of your house where they can give shade during the hot summer months. Also, install an energy efficient thermostat with a day and night timer."[2]

Reality-Testing Skeptics Say:

"Conservation groups and scientists have been making headlines in the past year, warning that shrinking sea ice could make wild [polar] bears extinct by the end of the century, possibly within just 20 years. Right now, though, Inuits like Nathaniel Kalluk here in Resolute Bay aren't exactly worried. 'There are a lot more bears now than before,' said Mr. Kalluk, who is 51 and has been hunting since childhood. 'We'll spot 20 to 30 bears on a hunting trip. Twenty years ago, sometimes we didn't see any at all. . . .'

"In Canada, home to most of the world's polar bears, the population has risen by more than 20 percent in the past decade. The chief reason for the rise is probably restrictions on hunting (for which conservationists deserve credit). . . . In the 1930s, the Arctic was as warm as it is now, and in the distant past it was even warmer."[3]

CO_2 HASN'T CONTROLLED EARTH'S PAST TEMPERATURES

Recently the CO_2 levels in the Earth's atmosphere and temperatures on the Earth's surface have both been rising. Does that mean that high CO_2 levels have been causing the Earth's warming? Or is it just coincidence?

According to the Greenhouse Theory, more CO_2 in the Earth's atmosphere will trap more of the Earth's own radiated heat, warming the lower atmosphere and ultimately the surface of the planet—all other things being equal. But the fact that the Earth's temperature has warmed only slightly since 1940, despite the huge clouds of greenhouse gases emitted from human activities, provides evidence that the human greenhouse effect must be so small that it presents little threat to the planet or its people. This is especially true if the current CO_2 levels have already used up almost all of that trace gas's ability to heat our planet. (Each additional increment of CO_2 causes less warming.)

Recent research has given us a much clearer picture of the global interaction between temperature and CO_2:

First, satellite and high-altitude weather balloon data confirm that the lower atmosphere is not trapping lots of additional heat due to higher CO_2 concentrations. It is hard to know how fast the Earth's highly variable surface is warming, but it is warming faster than the lower atmosphere where the CO_2 is accumulating. This is strong evidence that CO_2 is not the primary climate factor.

Second, the Antarctic ice cores tell us that the Earth's temperatures and CO_2 levels have tracked closely together through the last three ice ages and global warmings. However, CO_2 has been a lagging indicator, its concentrations rising about eight hundred years after the temperatures warm. This is additional evidence that CO_2 is not the forcing agent in recent global climate changes.

Oregon State Climatologist George Taylor recalls: "Early Vostok analysis looked at samples centuries apart, and concluded (correctly) that there is a very strong relationship between temperatures and CO_2 concentrations. The conclusion for many was obvious: when CO_2 goes up, temperatures go up, and vice-versa. This became the basis for a number of scary-looking graphs in books by scientist Stephen Schneider, former Vice President Al Gore, and others, predicting a much-warmer future (since most scientists agree that CO_2 will continue to go up for some time). Well, it's not as simple as that. When the Vostok data were analyzed for much shorter time periods (decades at a time rather than centuries), something quite different emerged. [Huburtus Fischer and his research team from the Scripps Institute of Oceanography] reported: '[T]he time lag of the rise in CO_2 concentrations with respect to temperature change is on the order of 400 to 1000 years.' *In other words, CO_2 changes are caused by temperature changes*"[4] (emphasis added).

According to Fischer, he "and his team analyzed the Vostok core going back 250,000 years, and cross-correlated their findings with a CO_2 record from the Antarctic's Taylor Dome covering the last 35,000 years—to get the temperature/CO_2 history in decades rather than centuries. They reported that 'The time lag of the rise in CO_2 concentrations with respect to temperature change is on the order of 400 to 1000 years during all three glacial-interglacial transitions.'"[5]

Fischer's team says the ocean gives up CO_2 when it and the atmosphere warm, which then stimulates more tree and plant growth on land. The trees and plants absorb CO_2, incorporating it into more and bigger roots and tree trunks, and more soil carbon sequestered under lush grasslands. The lag time of four hundred to one thousand years is related to the ocean-mixing time required for the CO_2 to be released from the water.[6]

Nicolas Caillon of the French Atomic Energy Commission used argon isotopes in the Antarctic ice cores to produce what he believed was an even more accurate record of the time lag for CO_2 increases after temperature increase—two hundred to eight hundred years. His conclusion: "This confirms that CO_2 is *not* the forcing that initially drives the climatic system during a deglaciation."[7]

FALLING TEMPERATURES IN THE POLAR REGIONS

Alarmists Lift Icy Fingers of Fear for the Ice:

"The warming will first be noticed in Arctic regions, for example in decreases in the sea ice extent. . . . Atmospheric warming will have far-reaching and dramatic consequences: for example, ocean circulation will be changed, the global sea ice cover will decrease, the distribution of precipitation will be different and the sea level will rise. Thus, everyone on the planet is participating in a completely uncontrolled, long-term, global experiment."[8]

"I slept under the midnight sun . . . on a 12-foot-thick slab of ice floating in the frigid Arctic Ocean. . . . But here, too, CO_2 levels are rising just as rapidly, and ultimately temperatures will rise with them—indeed *global warming is expected to push temperatures up much more rapidly in the polar regions . . .* since the polar cap plays such a crucial role in the world's weather system, the consequences of a thinning cap could be disastrous" (emphasis added).[9]

"UK scientists say parts of Antarctica have recently been warming much faster than most of the rest of the Earth. They believe the warming is probably without parallel for nearly two thousand years. . . . The authors say three of the four ice cores from the peninsula show a rise in temperature over the last half-century. And rapid regional warming has led to the loss of seven ice shelves during the last 50 years."[10]

Reality-Based Skeptics Still Question:

"A study in the *Journal of Climate* 13 (2000) finds that current trends in Antarctic sea ice are running in the opposite direction to the one predicted by climate models . . . based on data from the Defense Meteorological Satellite Program Special Sensor Microwave/Imager from Dec. 1987 to Dec. 1996, sea ice area and total sea ice extent has increased. . . . Another study appearing in the *Journal of Glaciology* 46 (2000) . . . found . . . it required 20,000 years for the ice sheet to fully respond to the different temperature changes. Scenarios of warming and cooling of 5 degrees C lead to a mere 1 to 1.5 percent change in initial ice sheet volume."[11]

"The authors point out a serious flaw in the IPCC's surface temperature record. . . . [T]here's virtually no information from Antarctica, which is known to have cooled slightly in recent decades. When the authors calculate the satellite-based temperature trend for the same regions actually covered by the IPCC, they find that the IPCC's geographic selection results in an overestimation of warming by 33%."[12]

It Should Start at the Poles

If the Greenhouse Theory were valid, temperatures in the Arctic and the Antarctic would have risen several degrees Celsius since 1940 due to the huge emissions of man-made CO_2 . The icy bad news for the CO_2 alarmists is that the temperatures at and near the North and South Poles are lower now than they were in 1930.

The Antarctic Peninsula, the thin finger of land pointing north toward Argentina (and the equator) has been getting warmer. We've heard an inordinate amount of hoopla about the warming on the peninsula, which makes up less than 3 percent of the Antarctic's land area. That's because (1) that is where most of the scientists and thermometers are; and (2) it is the only part showing any agreement with the Greenhouse Theory. The other 97 percent of Antarctica has been cooling since the mid-1960s.

The modern Antarctic network of long-term temperature measurements was established in 1957. Recently, a research team led by the University of Chicago's Peter Doran published a paper in *Nature* saying, "Although previous reports suggest slight recent continental warming, our spatial analysis of Antarctic meteorological data demonstrates a net cooling on the Antarctic continent between 1966 and 2000."[13]

The data from twenty-one Antarctic surface stations show an average continental decline of 0.008 degrees Celsius from 1978 to 1998, and the infrared data from satellites operating since 1979 show a decline of 0.42 degrees Celsius per decade.[14] David W. J. Thompson of Colorado State University and Susan Solomon of the National Oceanographic and Atmospheric Administration also report a cooling trend in the Antarctic interior.[15]

The sea ice surrounding the Antarctic continent also confirms cooling. Australia's A. B. Watkins and Ian Simmonds report increases in Southern Ocean sea ice parameters from 1978 to 1996 and an increase in the length of the sea-ice season in the 1990s.[16]

Alaska Is Just a Small Part of the Arctic

In the Arctic region, there's been much made of a warming trend in Alaska that may be a reflection of the Pacific Decadal Oscillation event in 1976–1977. Arctic-wide, however, there is no impending sign of warming or the ice cap melting.

Still, the Scaremongers Are Trying:

"There is emerging evidence that the Arctic permafrost, the soil that usually stays frozen all year round, is melting and releasing carbon. The Arctic contains about 14 percent of the carbon stored in the world's soils, and a release of the entire Arctic carbon store would add hugely to climate change. . . . The most imminent victims of the Arctic's melting permafrost are an estimated

200,000 indigenous people living in the Arctic region whose very existence is now under threat. . . . [D]amage, caused by the melting of permafrost, to buildings, roads, and pipelines is occurring in Alaska."[17]

"Anyone who doubts the gravity of global warming should ask Alaska's Eskimo, Indian, and Aleut elders. . . . [S]almon are increasingly susceptible to warm-water parasites and suffer from lesions and strange behavior. Salmon and moose meat have developed odd tastes, and the marrow in moose bones is weirdly runny, they say. Arctic pack ice is disappearing, making food scarce for sea animals and causing difficulties for the natives who hunt them. It is feared that polar bears, to name one species, may disappear from the Northern hemisphere by mid-century."[18]

Polish climatologist Rajmund Przybylak used the mean monthly readings from thirty-seven Arctic and seven sub-Arctic stations to construct the near-surface Arctic air temperatures over the last seventy years. He found "in the Arctic, the highest temperatures since the beginning of instrumental observations occurred clearly in the 1930s." Even the temperatures in the 1950s were higher than those of the last decade. In his second paper, which examined the temperatures for the entire Arctic for 1951 to 1990, he reported, "no tangible manifestations of the greenhouse effect [could] be identified."[19]

The University of Alaska's Igor V. Polyakov and his team analyzed the data from 125 Arctic land stations and a number of drifting buoys. They found a strong warming between 1917 and 1937, but no net warming—and perhaps a slight cooling—since 1937.[20]

Greenland has also been growing colder. Over the last half-century, there has been a statistically significant cooling, particularly in southwestern coastal Greenland. Sea surface temperatures in the nearby Labrador Sea have also fallen. The studies were made by Edward Hanna of Britain's University of Plymouth and John Capellan of the Danish Meteorological Institute using the data from eight Danish weather stations on Greenland, plus three stations collecting data from the surface of the nearby sea.[21] Apparently we can expect the Greenland Ice Sheet to last a few years longer.

WATER VAPOR—THE UNPROVEN ASSUMPTION

Water vapor is the most abundant and important greenhouse gas, even during the current warming. Water vapor makes up about 60 percent of the natural greenhouse effect, with CO_2 making up an estimated 20 percent. Such minor gases as ozone (O_3), nitrous oxide (N_2O), methane (CH_4), and several others make up the remaining 20 percent.

Despite all the talk about CO_2 and methane, most of the greenhouse effect's super-warming is *assumed* to come from water vapor. *Without this assumed large positive feedback from water vapor, the contribution of the other greenhouse gases would not produce a large temperature increase.*

The BBC's weather home page says: "The amount of water vapor in the atmosphere is not at all uniform—far from it—but changes drastically and abruptly, often in a matter of a few hours, to cause, for example, thunderstorms. It takes a lot of energy to evaporate water. A molecule of water vapor 'contains' much more energy than a molecule of liquid water. And quite a bit of water is evaporated every day as the Sun shines down on Earth's vast oceans. In short, water vapor is one of the most important 'storehouses' of energy in the atmosphere and in the climate system. . . . There is also substantial scientific uncertainty about water's role as vapor or clouds. Since it has both warming and cooling effects, water is a 'wild card.'"[22]

The water vapor assumption is based on a valid reality: that as the planet warms, more water evaporates into the air. Moreover, warmer air can hold more water. As an example, one kilo of air at $-15°C$. can retain only one gram of water, but at $35°C$. it can hold forty grams. But, as the BBC notes, that doesn't tell us if the extra water vapor is now staying up there to amplify the CO_2 warming—or raining back down more quickly.

A HUGE CLOUD-CONTROLLED
HEAT VENT OVER THE PACIFIC?

NASA reported in 2001 that MIT's Richard Lindzen and a NASA research team had found a huge climatic heat-vent over the warm pool of the Pacific, the planet's warmest spot. The vent seems to open naturally to release extra heat when the sea surface temperature rises.[23]

The team analyzed twenty months of detailed daily satellite observations of cloud cover and sea surface temperatures for the vast ocean region from Australia and Japan to the Hawaiian Islands. The cloud cover data came from Japan's GMS-5 Geostationary Meteorological Satellite and the sea surface temperatures came from the U.S. National Centers for Environmental Predictions.

"High clouds over the western tropical Pacific Ocean seem to systematically decrease when sea surface temperatures are higher," said Arthur Hou, a member of the Lindzen team. Those high clouds are the ones that trap heat in the atmosphere. "With warmer sea surface temperatures beneath the cloud, the coalescence process that produces precipitation becomes more efficient," said Lindzen. "More of the cloud droplets form raindrops and fewer are left in the cloud to form ice crystals. As a result, the area of cirrus cloud is reduced."[24] Icy cirrus clouds are poor sunshields, but very efficient insulators. A decrease in cirrus cloud area would cool the Earth by allowing more heat energy to leave the atmosphere.

The NASA-MIT study *strongly* suggests that the Earth is much more active in managing its atmospheric temperatures than the computer models have assumed, and thus much less sensitive to warming effects.

The Lindzen team's results confirmed an earlier study done by researchers from NASA's Goddard Space Flight Center. That team, led by NASA's Y. C. Sud, used satellite data on cloud cover and sea surface data from airplanes to analyze why the sea surface temperature in the "warm pool" varies almost entirely between 28 and 30°C.[25] They found that temperatures above 28° charge the air with more moisture, creating more clouds, even as they encourage cool, dry downdrafts of air that lower the sea surface temperature.

After the Lindzen paper was published, *Science* featured two more studies confirming the Lindzen heat vent findings on 1 February 2002. One was led by Junye Chen of Columbia University and the other by NASA's Bruce A. Weilicki.[26] Their common finding: that the Pacific's vertical heat vent emitted about as much heat energy during the 1980s and 1990s as is generally predicted for an instantaneous doubling of the air's CO_2 content. Yet only very slight changes in the sea surface temperature were observed.

NASA issued a press release on the discovery of the "heat vent." The original study was based on the best modern technology for climate observations: a geostationary satellite, with a massive accumulation of new data, and the most accurate sea surface temperatures taken since the ships threw away their galvanized buckets.

Aside from a few small think tanks, however, nobody paid any attention. Articles appear almost daily about "abrupt ice ages" and "millions of extinct species" based on nothing more than self-aggrandizing speculation. But the thrice-confirmed evidence that the Pacific Ocean tends to quietly and naturally vent its extra heat back to outer space—protecting the biosphere—isn't news.

Why not?

NOTES

1. World Wildlife Fund website, 2004, www.worldwildlife.org/climate/index.cfm

2. Johan Olsson, "The Effects of Global Warming," 12 January 1996, <http://www.geocities.com/TimesSquare/1848/global.html> (accessed April 2004).

3. John Tierney, "The Good News Bears," *New York Times*, 6 August 2005.

4. George Taylor, "Debate over Temperature Heats Up," 19 December, 2003, <www.techcentralstation.com> (June 2004).

5. H. Fischer et al., "Ice Core Record of Atmospheric CO_2 around the Last Three Glacial Terminations," *Science* 283 (1999): 1712–714.

6. Ibid.

7. N. Caillon et al., 2003, "Timing of Atmospheric CO_2 and Antarctic Temperature Changes across Termination III," *Science* 299 (2003): 1728–731.

8. The Nansen Environment and Remote Sensing Center, University of Bergen, March 1995, <http://www.bestofmaui.com/ournvmag.html> (accessed April 2004).

9. Al Gore, *Earth in the Balance* (New York: Houghton-Mifflin, 1992), 22–23.

10. BBC News, "Rapid Antarctic Warming Puzzle," 6 September 2001.

11. "Antarctica: To Melt or Not to Melt?" Competitive Enterprise Institute, Cooler Heads Project 5, no. 3 (7 February 2001).

12. "For Land's Sake," <www.worldclimatereport.com> (17 March 2004), commenting on A. T. J. de Laat and A. N. Maurellis, "Industrial CO2 Emissions as a Proxy for Anthropogenic Influence on Lower Tropospheric Temperature Trends," *Geophysical Research Letters* 31: l05204, doi: 10.1029/2003GL019024.

13. P. T. Doran et al., "Antarctic Climate Cooling and Terrestrial Ecosystem Response," *Nature Advance* 415 (2002): 517–20.

14. J. C. Comiso, "Variability and Trends in Antarctic Surface Temperatures from *in situ* and Satellite Infrared Measurements," *Journal of Climate* 13 (2000): 1674–696.

15. D. W. J. Thompson and S. Solomon, "Interpretation of Recent Southern Hemisphere Climate Change," *Science* 296 (2002): 895–99.

16. A. B. Watkins and I. Simmonds, "Current Trends in Antarctic Sea Ice: The 1990s Impact on a Short Climatology," *Journal of Climate* 13 (2000): 4441–451.

17. "The Climate Domino," *The Ecologist* 31, no. 3 (April 2001): 10.

18. "Global Warming Devastates Native Alaskans," Reuters News Service, 18 April 2004.

19. R. Przybylak, "Temporal and Spatial Variation of Surface Air Temperature over the Period of Instrumental Observations in the Arctic," *International Journal of Climatology* 20, (2000): 587–614; and Przybylak, "Changes in Seasonal and Annual High-Frequency Air Temperature Variability in the Arctic from 1951 to 1990," *International Journal of Climatology* 22 (2002): 1017–33.

20. I. V. Polyakov et al., "Variability and Trends of Air Temperature and Pressure in the Maritime Arctic, 1875–2000," *Journal of Climate* 16 (2003): 2067–77.

21. E. Hanna and J. Capellan, "Recent Cooling in Coastal Southern Greenland and Relation with the North Atlantic Oscillation," *Geophysical Research Letters* 30, (2003): 10.1029/2002GL015797.

22. *BBC Weather*, "Earth Gases—Water Vapour," February 2005, <http://www.bbc.co.uk/weather/features/gases_watervapour.shtml> (February 2005).

23. R. S. Lindzen et al., "Does the Earth Have an Adaptive Infrared Iris?" *Bulletin of the American Meteorological Society* 82 (2001): 417–32. Lindzen earned his Ph.D. from Harvard, participated in writing the IPCC report, and recently served on an eleven-member National Academy of Sciences panel evaluating climate science capabilities. He has won the American Meteorological Society's Meisinger and Charney Awards, and the American Geophysical Union's Macelwane Medal.

24. "Natural Heat Vent in Pacific Cloud Cover Could Diminish Greenhouse Warming," press release, American Meteorological Society, 28 February 2001.

25. Y. C. Sud et al., "Mechanism Regulating Sea-Surface Temperatures and Deep Convection in the Tropics," *Geophysical Research Letters* 26 (1999): 1019–22.

26. J. Chen et al., "Evidence for Strengthening of the Tropical General Circulation in the 1990s," *Science* 295 (2002): 838–41 and B. A. Weilicki et al., "Evidence for Large Decadal Variability in the Tropical Mean Radiative Energy Budget," *Science* 295 (2002): 841–44.

Chapter Four

The Baseless Fears

Sea Levels Will Surge,
Bringing Floods and Devastation

The Activist View of Sea Level Rise:

"The tiny Pacific nation of Tuvalu, population 11,000, is preparing to abandon ship in the face of rising sea levels. Tuvalu has applied for permission to move the people from nine low-lying coral atolls to Australia or New Zealand. The ocean level rose 20 to 30 centimeters during the 20th century and could rise one meter in the 21st, says Lester Brown, president of the Earth Policy Institute. The rise is due to thermal expansion of the ocean and melting of glaciers."[1]

"Previous studies suggest that the expected global warming from the greenhouse effect could raise sea level from 50 to 200 centimeters (2 to 7 feet) in the next century. . . . We estimate that if no measures are taken to hold back the sea, a one meter rise in sea level would inundate 14,000 square miles [of the United States]. . . . The 1,500 square miles (600–700 square miles) of densely developed coastal lowlands could be protected for approximately one to two thousand dollars per year for a typical coastal lot. . . . Although . . . pumping sand would allow us to keep our beaches, levees and bulkheads along sheltered waters would gradually eliminate most of the nation's wetland shorelines. . . . To ensure the long-term survival of coastal wetlands, federal and state environmental agencies should begin to lay the groundwork for a gradual abandonment of coastal lowlands."[2]

A Professional Comments on Sea Level Rise:

"In the past 150 years, sea level has risen at a rate of 6 inches (plus or minus 4 inches) per century and is apparently not accelerating. Sea level also rose in the 17th and 18th centuries, obviously due to natural causes, but not as much. Sea level has been rising naturally for thousands of years (about 2 in. per century in the past 6,000 years). If we look at ice volumes of past interglacial periods and realize how [slowly] ice responds to climate, we know that in the current interglacial period (which began about 11,000 years ago) there is still more land ice available for melting, implying continued sea level rise with or without climate change.

"One of my duties in the office of the State Climatologist is to inform developers and industries of the potential climate risks and rewards in Alabama. I am very frank in pointing out the dangers of beachfront property along the Gulf Coast. A sea level rise of 6 in. over 100 years, or even 50 years, is miniscule compared with the storm surge of a powerful hurricane like Fredrick or Camille. Coastal areas threatened today will be threatened in the future. The sea level rise, which will continue, will be very slow and thus give decades of opportunity for adaptation, if one is able to survive the storms.

"The main point I stress, to state and local agencies as well as industries, is that they invest today in infrastructure that can withstand the severe weather events that we know are going to continue. These investments include extending floodway easements, improvement in storm water drainage systems and avoiding hurricane-prone coastal development, among other actions."[3]

WILL A SEA LEVEL RISE OF FOUR INCHES BY 2100 DESTROY CIVILIZATION?

The United Nations' Intergovernmental Panel on Climate Change in 1990 predicted that man-made warming would produce a sea level rise of thirty to one hundred centimeters by 2100.[4] By 2001, the IPCC's *Third Assessment Report* had lowered its predicted sea level increase slightly to nine to eighty-eight centimeters.[5] That's still a potentially massive sea level rise. However, it reveals an even more massive uncertainty: a ten-fold range of doubt.

The IPCC has in fact been harshly criticized for its handling of sea level issues by the International Union for Quaternary Research. INQUA is a seventy-five-year-old scientific organization dedicated to researching global environmental and climatic changes over the past 2 million years.[6] INQUA's Commission on Sea Level Changes and Coastal Evolution says that the IPCC has ignored the scientists who produced most of the data and observations in sea level science, substituting unverified model results instead.

Nils Axel Morner, the Swedish geologist who was president of the Sea Level Commission, says, "This is nothing but falsification of scientific observational facts."[7] Morner says sea level shows no trend at all over the past three hundred years, and satellite telemetry shows virtually no change in the past decade. This is contrary to the model predictions of the IPCC. "This implies that there is no fear of any massive future flooding as claimed in most global warming scenarios," says Morner.[8]

The IPCC offers a range of sea level rise of "0.09 to 0.88 meters between 1990 and 2100." The Sea Level Commission's expert-based figure is "10 cm—plus or minus 10 cm." In other words, the sea level scientists believe there is no way to scientifically predict any sea level rise at all in the twenty-first century.

The U.S. Environmental Protection Agency (EPA) itself has published a study that says global sea level has a 50 percent chance of rising 45 cm (1.5 feet) by the year 2100—and only a 1 percent chance of rising 110 cm (over 3.5 feet).[9] Presenting these odds is a defensible position scientifically— but the EPA knew the newspapers would write it as: "The EPA says sea levels may rise as much as 3.5 feet, in line with the warnings of the IPCC."

Neither the IPCC nor the EPA is offering honest assessments of the very modest increases in sea level (4–6 inches) that are most likely in the twenty-first century. We have no reason to expect a big sea level rise in the next century, or the one after that.

Nor will it be necessary to "armor" the shorelines and destroy coastal wetlands.

WHY DO THE IPCC AND THE EPA WARN OF BIG SEA LEVEL INCREASES?

Global warming advocates seem to assume a huge increase in sea level is immediate and inevitable if the planet continues to warm. Sea level rise is a product of conflicting forces, however.

Warmer temperatures expand the volume of the water. Warmer temperatures melt more glacier ice. But warmer temperatures also evaporate more water from the oceans and lakes. When the clouds deposit the increased moisture from that rapid evaporation on polar ice caps and glaciers around the world, *the ice caps and glaciers will actually grow* unless the local temperatures are warm enough to increase local melting.

Time is also a critical factor. Ice melts slowly. Glaciers and ice caps can take thousands of years to melt completely because their surfaces reflect away so much of the sun's heat.

That's why the West Antarctic ice sheet, at least 10,000 years past its last ice age, still has another 7,000 years worth of ice to melt, according to John Stone of the University of Washington.[10] Stone and his team analyzed the chemical composition of the rocks left behind on the mountains of Antarctica's Ford Range when the ice began to retreat.

Given the Earth's highly variable climate history, another cooling period is almost certain to intervene before the West Antarctic Ice Sheet disappears.

Walter Munk of the Scripps Institute of Oceanography reports that glacial melting due to higher twentieth-century temperatures can account for only four inches of sea level rise or fall per century. Essentially, we do not know why sea levels in recent decades have sometimes risen at double that rate— but the average is six inches per century.[11]

The tectonically stable coastline of the Chukchi Sea in northwest Alaska shows sea levels there have risen only about a quarter of a millimeter per year over the past 6,000 years. However, there have been several periods of both slower and more rapid increases.[12]

The world's longest set of sea-level observations has been faithfully recorded for more than a thousand years at Stockholm, Sweden. According to M. Ekman, they tell us that "sea level changes due to northern hemisphere climate variations since A.D. 800 have probably always kept within –1.5 and +1.5 mm/yr, with an average fairly close to zero."[13] This result comes about because land surfaces there are rising, thus offsetting the rise in ocean levels. The reason is the rebounding of the land as it adjusts to the removal of the loads from the ice sheets that had covered it earlier.

Niels Reeh of the University of Denmark has reported a "broad consensus" among sea level experts that another 1 degree Celsius of warming would create only a tiny change in global sea levels. He says the melting of Greenland's ice sheet would increase sea level by only 0.3 to 0.77 mm per year. Meanwhile, Antarctica would *subtract* 0.2 to 0.7 mm per year as increased precipitation added to its ice cap.[14]

WHAT IS SINKING AND WHAT ISN'T?

Will rising seas *submerge islands*? Let's answer this question by looking at the Maldive Islands. The Maldives, 1,200 low-lying islands in the Indian Ocean, currently sit only one or two meters above sea level. Their 300,000 people and infrastructure are so exposed that INQUA made the Maldives a priority research target.

They found that the sea level around the Maldives has been "oscillating" over the past 5,000 years, rising and falling in response to short-term local events.

There have been four periods when the sea level was higher than today: About 3,900 years ago, the ocean was 1.1 to 1.2 meters above current sea level. (The very warm Climate Optimum ended about 5,000 years ago.) Then, 2,700 years ago, the Indian Ocean was just a bit higher than today, at 0.1 to 0.2 meters above current levels. At the peak of the Medieval Warming (950–1300), the sea was 0.5 to 0.6 meters higher than today.

There are clear indications of a recent fall in the Maldives sea level. Nils Axel Morner himself led a research team that concluded that the sea level "experienced a general fall of the order of 20–30 cm" since 1970.[15] The most surprising finding was that the Indian Ocean was higher than today between 1900 and 1970, and that sea level has fallen since then—in the middle of a warming period. The team sees no reason, either historically or currently, to expect the Maldives to become flooded in the near future.

Sinking Tuvalu?

Lester R. Brown is an agricultural economist who has been wrongly predicting massive world famines since the 1960s—during a period when the high-yield rice and wheat crops of the Green Revolution were sharply increasing per capita world food production.

In 2001, Brown claimed that the First World was unfairly flooding the Pacific islands through global warming caused by the greenhouse gases of its power plants.[16] Since Brown has a history of making unsubstantiated claims, let's look more closely at Tuvalu.

Geologically, Tuvalu is in a risky position. Its atolls rest on volcanic rock that is gradually subsiding into the sea. On top of the sinking rock is coral, which grows only slowly, and whose lower layers die off as they sink too far into the ocean for the corals to get sunlight. It does seem unlikely that Tuvalu could survive the rapid sea level rise forecast by the activists.

In 2004, Brown said that the oceans rose eight to twelve inches in the last century and could rise another eighteen inches by 2100.

Since 1993, however, the TOPEX/POSEIDON satellite radars found that Tuvalu's sea levels have *fallen* four inches over a decade.

A second, independent record confirms the satellite measurement: the modern tide gauges installed on Tuvalu in 1978. They show that a strong El Niño in 1997–1998 dropped Tuvalu's sea level by about one foot. The El Niño/Southern Oscillation is a natural periodic phenomenon that makes no difference at all in long-term sea level trends.

The director of Australia's National Tidal Facility, Wolfgang Scherer, says, *"One definitive statement we can make is that there is no indication based on observations that sea level rise is accelerating"*[17] (emphasis added).

If the sea around Tuvalu is not rising, why is Tuvalu's prime minister upset?

First, some Tuvaluans may actually hope that filing international "lawsuits" will deliver big chunks of cash to them as "reparations" for global warming.

Second, if they were allowed to move their 11,000 residents to the beaches of affluent countries like Australia or New Zealand, they would expect to prosper. Tuvalu is a small, remote place with limited fresh water and limited job opportunities.

On the other hand, an environmental official on the island, Paani Laupepa, says that the islanders have been excavating sand for building projects, which has given the impression to casual observers that the sea has risen. "The island is full of holes and seawater is coming through these, flooding areas that weren't normally flooded 10 or 15 years ago," says Laupepa.[18]

Another environmental official on Tuvalu, Elisala Pita, in an interview with the *Toronto Globe and Mail* (24 November 2001), said, "Tuvalu is being used for the issue of climate change. People are telling all these lies, just using Tuvalu to prove their point. No island is sinking. Tuvalu is not sinking."

Is Venice Drowning?

What about Venice? The whole world knows that the priceless architectural treasures of the Italian canal city are threatened by rising water levels. Doesn't that prove the rising sea levels are a problem? Venice is now flooded about forty-three times a year, compared with seven floods per year a hundred years ago.

The Italian National Research Council says the relative sea level in Venice rose twenty-three centimeters between 1897 and 1983. But twelve centimeters of the apparent sea level rise was due to land subsiding in and around Venice, much of it due to the weight of buildings and bridges on the soft soils of the coastal region. The city is erecting mobile barriers and internal water defense structures, to minimize damage from high water levels during high tides. They are the best ways we have today to adapt to the apparently inevitable sinking of the Venetian land mass.

LONGER-TERM SEA LEVEL CHANGE

D. C. "Bruce" Douglas of the Hurricane Center at Florida International University and W. Richard Peltier of the University of Toronto were recently asked to address the broader question of long-term change in sea level for *Physics Today*. They noted that today's changes in sea level are tiny compared with those of past ice ages and warming periods. Ancient corals found in Barbados reveal that sea level increased by about 120 meters (394 feet) since the last ice age began melting some 21,000 years ago. By about 5,000–6,000 years ago, most of the ice age's trillions of tons of extra ice had melted.[19] After that, global sea level rise slowed, and apparently stabilized about 3,000 to 4,000 thousand years ago. No studies have detected any significant acceleration during the twentieth century.

LOOKING TO THE FUTURE AND PREPARING FOR IT

A continued slow rise in sea levels is probably the biggest real danger of a moderate global warming, but it's a far smaller problem than the activists have painted it. Six inches per century is slow, but if it continues over the next five hundred years, it could add up to significant coastal changes.

The world wouldn't lose its wetlands due to rising sea levels. The wetlands and their species would simply move slightly upslope, as they have so many times in the past.

More than one observer has declared that we would have to "build a dike around Bangladesh," to prevent higher sea levels from destroying that low-lying country and drowning millions of people. Actually, building a dike around Bangladesh may not be a bad idea anyway.

The problem is not sea levels per se, but the storm surges from huge tropical cyclones, which have hit the country every three to four years in recent decades. Bangladesh has built a large number of "typhoon towers," that allow the population to climb above the floods with the belongings they can carry. However, the floods of salt water often stay for weeks, spreading disease, poisoning soils, halting economic activity, and inflicting massive amounts of physical damage on buildings, roads, bridges, water systems, and so forth.

A dike would be expensive. It would have to be done with great sensitivity to the coastal wetlands. But perhaps it will need to be done.

Even with a slow and moderate rise in sea level, it would also make sense to install tougher zoning for low-lying areas and tougher building codes for areas within reach of storm surges. The massive problems inflicted by Hurricane Katrina on the city of New Orleans and the communities of the Gulf Coast in 2005 underscore that point, and remind us that normal hurricane risks are amplified as our cities grow and our people seek waterfronts to live and play on. At the very least, America should stop encouraging high-risk waterside building through government-financed flood insurance.

What about low-lying islands? Since there are only about 11,000 Tuvaluans, emigration does not present insurmountable problems—though it would be regrettable if that became necessary. And, the Tuvaluans might even prefer it.

The point here is that we live on a planet where the climate has been changing constantly for the past billion years. There will always be an ebb and flow between land and water, and their interface will be rich with competing organisms. We expect a continued abundance of the corals, coastal forests, and horseshoe crabs—not to mention sand fleas, biting flies, mosquitoes, and beachcombers.

NOTES

1. "Global warming claims first victim," University of Wisconsin, 20 November 2001, <http://whyfiles.org/update/091beach> (accessed February 2004).

2. James Titus et al., "Greenhouse Effect and Sea Level Rise: The Cost of Holding Back the Sea," *Coastal Management* 19 (1991): 171–204.

3. John Christy, Alabama State Climatologist, before the U.S. House Committee on Resources, 13 May 2003.

4. R. A. Warwick and J. Oerlemans, "Sea Level Rise," in *Climate Change, The IPCC Assessment*, J. H. Houghton, G. J. Jenkins, and J. J. Ephron, eds. (Cambridge, UK: Cambridge University Press, 1990).

5. Intergovernmental Panel on Climate Change, *Third Assessment Report* (Cambridge, UK: Cambridge University Press, 2001).

6. The International Union for Quaternary Research may be reached through its Secretary-General, Peter Coxon, Department of Geography, Trinity College, Dublin, Ireland, <pcoxon@tcd.ie>.

7. Nils Axel Morner, letter to Cambridge Conference Network, 27 April 2001, abob.libs.uga.edu/bobk/ccc/cc042701.html.

8. Nils Axel Morner, "Estimating Future Sea Level Changes from Past Records," *Global and Planetary Change* 40, issues 1–2 (January 2004): 49–54.

9. U.S. Environmental Protection Agency, *Global Warming—Climate, Sea Level*, <http://yosemite.epa.gov/oar/globalwarming/nsf/content/ClimateFutureClimateSea.level.html>.

10. J. Stone et al., "Holocene Deglaciation of Marie Byrd Land, West Antarctica," *Science* 299 (2003): 99–102.

11. W. Munk, "Ocean Freshening, Sea Level Rising," *Science* 300 (2003): 2014–43.

12. O. W. Mason and J. W. Jordan, "Minimal Late Holocene Sea Level Rise in the Chukchi Sea: Arctic Insensitivity to Global Change?" *Global and Planetary Changes* 32 (2002): 13–23.

13. M. Ekman, "Climate Changes Detected through the World's Longest Sea Level Series," *Global and Planetary Change* 21 (1999): 1215–224.

14. N. Reeh, "Mass Balance of the Greenland Ice Sheet: Can Modern Observation Methods Reduce the Uncertainty?" *Geografiska annaler* 81A (1999): 735–42.

15. Morner et al., "New Perspectives for the Future of the Maldives," *Global and Planetary Change* 40 (2004): 177–82.

16. Lester R. Brown, "Rising Sea Level Forcing Evacuation of Island Country," press release, Earth Policy Institute, 15 November 2001.

17. Quoted by Sallie Baliunas and Willie Soon, "Lester's Brownout: Activist Exploits Poor Islanders," 10 December 2001, <www.techcentralstation> (June 2004).

18. S. Baliunas and W. Soon, "Is Tuvalu Really Sinking?" February 2002, www.pacificislands.cc.

19. D. C. Douglas and W. R. Peltier, "The Puzzle of Global Sea-Level Rise," *Physics Today* 55 (March 2002): 35–40.

Chapter Five

The Treaty that Would Change
Earth's Climate—Or Maybe Not

The Kyoto Supporters and How They Grew:

"Climate Action Network Europe is a worldwide network of over 365 Non-governmental Organizations (NGOs) working to promote government, private sector, and individual action to limit human-induced climate change to ecologically sustainable levels. CAN is based on trust, openness and democracy . . . receiving funding in the present financial year from the European Commission, the Dutch Government and the Belgian Government."[1]

"The U.S. Climate Action Network (USCAN). . . . Made up of over forty environment, development, and energy nongovernmental organizations, the network works to inform and affect U.S. domestic and international policies. U.S. NGOs have highly professional staffs with well-developed climate and energy policy programs, setting the stage for their heavy involvement in climate and energy policies . . . within the UN."[2]

Man-Made Warming Skeptics Answer:

"A suppressed report by the federal government [of Canada] evaluating the effectiveness of spending $500 million since the year 2000 to reduce emissions of greenhouse gases has shown—surprise!—that the spending was largely wasted, producing neither a reduction in greenhouse gas emissions nor the development of new 'cleaner' technologies. An anonymous source that participated in the mid-term review is quoted in the *Star*, saying, 'We seriously underestimated the difficulty of getting reductions and overestimated the payoff from new technologies. . . . Action Plan 2000 . . . committed $210 million to promote technologies that reduced greenhouse gas emissions . . . [and] $125 million to cities to encourage them to use the nonexistent new technologies. And another $100 million was spent on promoting foreign demand for the nonexistent new technologies.'"3

The Kyoto Protocol was produced by an alliance between environmental organizations and the appointed functionaries of the United Nations. Neither group was elected to anything nor did either group control any people or territory.

They nevertheless used the generally favorable public attitudes toward environmental conservation to demand what they called an "insurance policy for the planet" against man-made overheating from CO_2. The NGOs used their new computers and the nascent Internet to organize one of the most impressive volunteer efforts in modern times. Nearly 20,000 environmental activists went to Brazil in 1992 for the "Earth Summit." When that depth of interest became evident, governments rushed to announce their official delegates. More than 170 governments were represented, a startling 108 by their "heads of state."

Most of the hordes of activists actually attended a parallel "cheerleaders" conference called the Nongovernmental Organization Forum, which was held nearby. However, 2,400 activists were official delegates to the summit itself. They were highly organized, and constantly referred to the huge numbers of their colleagues meeting across town and waiting for "action on behalf of the planet."

Politicians naturally saw the world's hundreds of thousands of earnest and energetic environmental activists as a movement to be co-opted. European politicians were especially eager, since the Green parties there were often key parts of their fragile governing coalitions, or soon likely to be. They wanted something to give the Greens that wouldn't cost money before the next election.

THE GREEN DREAM

What the Greens wanted was to end or severely restrict the use of fossil fuels. This was the era in which biologist Paul Ehrlich wrote scathingly that the global problem was "too many rich people."4 The activist movement saw rich people as the ones using too many resources. They saw cheap energy as the root cause of the technological abundance underlying the "throw-away society." In its turn, cheap energy produced too many rich people and enticed poor people with the idea that they could get rich, too.

The Greens wanted solar and wind energy to be appreciated, never mind that as energy sources they were expensive and erratic. They believed that high-yield farming was causing overpopulation by feeding too many people—and high-yield farming depended on industrial nitrogen fertilizer, which is produced with the use of fossil fuels. They ardently demanded only organic farming, with half the yield per acre and radically less capacity to support population growth.

There was no real evidence that fossil fuels were overheating the world, then or now. Theory says that more greenhouse gases in the atmosphere will trap more heat, but no one knows whether the amounts of heat trapped by CO_2 increases are significant. Nothing in the Earth's climate history confirms CO_2 as a strong driver of climate warming.

It was certainly true, however, that nothing would disarm modern technologies quite so completely as depriving the First World of energy. Automobiles would be abandoned. The impetus for locally grown food would become irresistible. It would have to be organic food, too, because without fossil fuels to run the fertilizer factories, there'd be no industrial plant food. Without fertilizer, however, millions of humans might starve, even as more forest was cleared for low-yield crops. There is no need, however, to harshly "starve out" human population growth. Birth rates are already plummeting worldwide and, thus, human numbers are set to begin a long, slow decline after 2050.

The United Nations, for its part, saw the Greenhouse Theory as a way of expanding its influence and power. The Greenhouse Theory demanded that energy be scarce, and the agency that rationed out the energy would be powerful indeed.

The Great Atmospheric Science Funding Balloon

But wait! Wouldn't the scientists blow the whistle on a fraudulent "Global Warming"? Wouldn't scientific peer review protect the public from a misguided belief in human responsibility for Earth's natural climate changes?

Apparently not. Much of the news about climate change is being produced by highly qualified scientists, using multimillion-dollar supercomputers to run billion-dollar coupled global circulation models projecting unverified climate guesses a hundred years into the future—where any mistakes will be hugely amplified. Other scientists are putting up highly sophisticated satellites, running long-range aircraft to monitor atmospheric trends, packing research ships with equipment for voyages to the Antarctic—and writing scary press releases to keep the research grants flowing.

Most of the funding to date has gone to atmospheric research, but marine scientists have recently begun demanding their share, claiming that global

Let's sneak a quick look back at the scientific community's reaction to the "global cooling scare" of the early 1970s.

"SCIENTISTS ADD TO HEAT OVER GLOBAL WARMING"

Excerpted from a column by S. Fred Singer, *Washington Times*, 5 May 1998.

"All that the [Council of the National Academy of Sciences] musters in defense of the Kyoto Accord is a scientifically outdated NAS/National Research Council report titled *Policy Implications of Greenhouse Warming*. Quoting from this 1991 report, the Council concludes that 'even given the considerable uncertainties in our knowledge of the relevant phenomena, greenhouse warming poses a potential threat sufficient to merit prompt responses. . . .'

"[I]n 1975, the NAS 'experts' exhibited the same hysterical fears—this time, however, asserting a 'finite possibility that a serious worldwide cooling could befall the Earth within the next 100 years. . . .'

"In *The Cooling: Has the Next Ice Age Already Begun? Can We Survive It?* published in 1975 by Prentice-Hall. Its author, Lowell Ponte, captures the then-prevailing mood: 'The NAS report was shocking, for it represented a warning from some of the world's most conservative scientists that an Ice Age, beginning in the near future . . . was not impossible.' *Contending that we may be 'on the brink of a [100,000 year] period of colder climate,' the NAS urged an immediate near-quadrupling of funds for research.* 'We simply cannot afford to be unprepared for either a natural or man-made climatic catastrophe" (emphasis added).

"Mr. Ponte lectures the public: 'Global cooling presents humankind with the most important social, political, and adaptive challenge we have had to deal with for 110,000 years. Your stake in the decisions we make concerning it is of ultimate importance: the survival of ourselves, our children, our species.'

"Any of this sound familiar?"

warming will bring on "abrupt global cooling" by shutting down the oceans' conveyor currents.

WHAT EXACTLY IS KYOTO?

The Kyoto Protocol is an international agreement that was ostensibly intended to limit the use of fossil-based energy on Earth. However, it was never likely to stabilize (let alone reduce) the world's greenhouse emissions.

The U.S. Senate declared in advance it would not ratify such a treaty, and never has.

The Kyoto Protocol derives from the Framework Convention on Climate Change (FCCC). The Convention on Climate Change is itself a treaty concluded at the 1992 Earth Summit in Rio.

Article 2 of the FCCC states that its ultimate objective is to "achieve stabilization of greenhouse gas concentration in the atmosphere at a level that would prevent dangerous anthropogenic interference with the climate system." Nowhere, in either the FCCC or in the Kyoto Protocol, is there any statement of what greenhouse gas levels might be "dangerous" to either humans or the environment. Or how.

The Kyoto treaty was negotiated by the Clinton administration in 1997, much of it personally by Vice President Al Gore in preparation for his unsuccessful run for the U.S. presidency in 2000. However, the Clinton–Gore administration never dared to bring the treaty to a Senate vote.

The U.S. Senate had already passed the Byrd–Hagel resolution on 21 July 1997 during the run-up to the UN's Kyoto meeting—voting 95 to 0 against any such treaty.[5] The resolution said that any climate treaty that did not include the developing countries was "inconsistent with the need for global action on climate change and is environmentally flawed." The Senate resolution also pointed out that if such a treaty left out the Third World, then "[t]he level of required emission reductions would result in serious harm to the U.S. economy, including significant job loss, trade disadvantages, increased energy and consumer costs, or any combination thereof."[6]

The completed Kyoto Protocol confirmed the fears of the U.S. Senate. It did not include the big developing countries and it did propose to put the full burden of climate protection on the United States and other First World countries. The reason was simple. The Third World was much more afraid of being left in poverty than it was of the largely benign climate trend revealed by the First World's thermometers. If Kyoto had required the signatures of China and India, it would never have been concluded. The UN's "evidence" for human-induced warming was essentially limited to repeating the mantra that "the Earth has warmed 0.6° C. in the last century," reciting the Greenhouse Theory, and offering printouts from complex but unverified computer models.

The Kyoto treaty was particularly attractive to the European governments that have taxed energy heavily for decades. For competitive reasons, Europe wanted very much to see the United States and its famous job-creating economy saddled with the same high energy costs that European employers and drivers already paid. (A barrel of oil that has netted the Saudi oil industry $35 has often yielded the British government $150 in taxes—with the taxes sanctified to "save the planet.")

The first stage of Kyoto energy constraints called for its members to reduce their emissions of greenhouse gases by a very modest 5.2 percent from 1990 levels. This would make only an undetectable impact on any global warming caused by human-generated CO_2. Even supporters of Kyoto admit that it would reduce calculated temperatures by only 0.05 degrees Celsius by 2050.[7]

The reason environmental groups were excited about Kyoto was the second, severe phase of the treaty—which is supposed to kick in as of 2012. The second phase envisioned much tighter constraints on greenhouse gas emissions that would begin to stabilize global greenhouse gas emissions, though the next level of greenhouse gas reduction has never been negotiated.

Indeed, the IPCC's *First Assessment Report* in 1990 stated that global fossil fuel use would have to be reduced by 60 to 80 percent to stabilize CO_2 levels.[8]

The United States would be required to reduce its greenhouse gas emissions by 7 percent from 1990 levels in the first phase of Kyoto. Dr. Harlan Watson, the senior U.S. negotiator at the 2004 Conference of the Parties (COP) meeting in Buenos Aires, reported that the United States will emit about 16 percent more greenhouse gases in 2010 than it did in 1990. In order to meet the Kyoto targets, the United States would thus have to cut its projected emissions by 23 percent by 2008. Since fossil fuels still provide about 85 percent of American energy, the United States would have to cut its energy use by nearly one-fourth unless it could rapidly and radically increase supplies of nuclear, wind, and solar power.

In the second phase of Kyoto, if all protocol member nations were required to cut fossil fuels by 60 percent from 1990 levels, America would probably be required to eliminate virtually all fossil fuel use while poor countries expanded their use.

The Protocol was stalled from its signing in 1997 until 2005, because its signing nations did not represent 55 percent of the global greenhouse gas emissions. The treaty was finally rescued by Russia, which ratified the treaty in late 2004, bringing the membership up to the required percentage. The Kyoto Protocol was implemented on 16 February 2005.

We still do not know how difficult it will be to cut back fossil fuel use in the real world. Few Kyoto member nations have yet actually attempted to cut their real-world greenhouse gas emissions. The choice of 1990 as the base year gave major advantages to Britain, Germany, and Russia. Britain had shut down its antique coal mines and shifted heavily to cleaner North Sea natural gas. Germany got Kyoto credit for shutting down the dirty industries built by East Germany's communist government. Russia got most of the credit for eliminating the former Soviet Union's heavily polluting factories. As of 2005, only Sweden and Britain, of all the European Union members, were on target to meet their 2008 Kyoto commitments.

At the same time that Russia joined Kyoto and enabled its implementation, Russian membership also virtually guaranteed that Kyoto would not actually reduce its members' emissions of greenhouse gases. Russia is expected to sell billions of dollars worth of its Soviet-era emission credits to the European countries, which will allow Europe to meet Kyoto's first-phase requirements without actually cutting emissions or energy use.

Yale University economist William Nordhaus has estimated that first-phase Kyoto emissions reductions would cost $716 billion, and that the United States would bear two-thirds of the global costs, which may be as good a guess as anyone's. No one has even tried to estimate the cost of fully stabilizing human-generated CO_2 emissions without some major technical breakthrough.

The U.S. electorate remains split 50–50 between a party that opposes Kyoto and a party that essentially favors it. (Democratic presidential candidate John Kerry said in a 2004 debate with President Bush that he would alter Kyoto so that the United States could sign it.)

THE UNITED NATIONS PANEL'S
FALSE EVIDENCE OF MAN-MADE WARMING

One of the most serious problems with the Kyoto treaty is that it has been promoted by essentially false statements from the UN Intergovernmental Panel on Climate Change.

The Sulfate Aerosol Fraud

The IPCC's 1990 report claimed that computerized global climate models showed a warming trend "broadly consistent" with real-world observations. But the real-world observations were of a slow, erratic global warming that started too early to be blamed on human CO_2 emissions. It has been going on in inexplicable surges (1850–1870, 1920–1940) instead of following the strong, steady upward trend of human CO_2 emissions after 1940. Moreover, the IPCC and the models were embarrassed about the 1940–1975 cooling trend, which no one had predicted, and which the IPCC could not explain.

To cover itself, the IPCC added a cooling factor to the greenhouse analysis after 1990—a claim that tiny aerosol particles produced by the emissions of sulfur dioxide from electric power plants had overwhelmed the warming effect of the CO_2.

The second IPCC report, published in 1996, again invoked the sulfate aerosol effect and produced the memorable but essentially meaningless phrase in its summary that "the balance of evidence *suggests* a human effect on climate"[9] (emphasis added).

By the time the third IPCC report was published in 2001, the sulfate aerosol "fix" had proved in conflict with observed reality.

The aerosols are produced mainly where industrial activity is highest. Therefore, the northern hemisphere should warm more slowly than the Southern hemisphere, since the sulfates produced there would reflect some sunlight, reduce incoming energy, and thereby offset part of the calculated greenhouse warming. But observations showed exactly the opposite: The highest rate of warming in the most recent twenty-five years had occurred at northern mid-latitudes, just where the most aerosols are emitted.

The third IPCC report swept the aerosol question under the rug, thus removing the factor that had "enabled" it to say its model observations were consistent with reality. However, the 2001 report kept its preconceived conclusion that "new evidence" made it likely that "most of the warming of the past 50 years" came from the human production of greenhouse gases.[10]

"Discernible Human Influence on Climate" Never Documented

Climate is so complex and variable that it's difficult to distinguish the causes of its variations. The technique adopted by the IPCC for its 1996 report was called "fingerprinting." The IPCC compared the detailed geographic patterns of climate change with the calculations of the climate models. This comparison, as published in the IPCC's Second Assessment in 1996, seemed to indicate a growing correspondence between real-world observation and modeled patterns.

On examination, however, this result proved to be false. The correspondence is produced only for the time interval 1943 to 1970. More recent decades show no such correspondence. Nor does the complete record, which dated from 1905 to 1995. The IPCC claim is based on selective data. Under the rules of science, this cancels the IPCC's claim of having found a human impact on climate.

The IPCC's defenders claim that the crucial chapter 8 of the panel's 1996 report was based on 130 peer-reviewed science studies. Actually, the chapter was based mainly on two research papers by its lead author, Ben Santer, of the U.S. government's Lawrence Livermore National Laboratory. Neither of the Santer papers had been published at the time the chapter was under review and had not been subject to peer review. Scientific reviewers subsequently learned that both the Santer papers shared the same defect as the IPCC's chapter 8: Their "linear upward trend" occurs only from 1943 to 1970.

In fact, *the IPCC report itself documented the reality that the human-made warming claim was false*. The "fingerprint test," as displayed in figure 8.10b of the 1996 report, shows the pattern correlation between observations and climate models *decreasing* during the major surge of surface temperature warming that occurred between 1920 and 1940.

The IPCC's 1996 report was reviewed by its consulting scientists in late 1995. The "Summary for Policy Makers" was approved in December, and the full report, including chapter 8, was accepted. However, after the printed report appeared in May 1996, the scientific reviewers discovered that major changes had been made "in the back room" after they had signed off on the science chapter's contents. Santer, despite the shortcomings of the scientific evidence, had inserted a strong endorsement of man-made warming in the 1996 report's chapter 8:

> There is evidence of an emerging pattern of climate response to forcing by greenhouse gases and sulfate aerosols . . . from the geographical, seasonal and vertical patterns of temperature change. . . . These results point toward a human influence on global climate.[11]

Santer added the following sentence to the crucial chapter 8 (of which he was the IPCC-appointed lead author) of the printed version of the 1996 IPCC report:

> The body of statistical evidence in chapter 8, when examined in the context of our physical understanding of the climate system, now points to a discernible human influence on the global climate.[12]

Santer also *deleted* these key statements from the expert-approved chapter 8 draft:

- "None of the studies cited above has shown clear evidence that we can attribute the observed [climate] changes to the specific cause of increases in greenhouse gases."
- "While some of the pattern-base studies discussed here have claimed detection of a significant climate change, no study to date has positively attributed all or part [of the climate change observed] to [man-made] causes. Nor has any study quantified the magnitude of a greenhouse gas effect or aerosol effect in the observed data—an issue of primary relevance to policy makers."

- "Any claims of positive detection and attribution of significant climate change are likely to remain controversial until uncertainties in the total natural variability of the climate system are reduced."
- "While none of these studies has specifically considered the attribution issue, they often draw some attribution conclusions, for which there is little justification."
- "When will an anthropogenic effect on climate be identified? It is not surprising that the best answer to this question is, 'We do not know.'"

Santer single-handedly reversed the "climate science" of the whole IPCC report—and with it the global warming political process! The "discernible human influence" supposedly revealed by the IPCC has been cited thousands of times since in media around the world, and has been the "stopper" in millions of debates among nonscientists.

The journal *Nature* mildly chided the IPCC for redoing chapter 8 to "ensure that it conformed" to the report's politically correct Summary for Policymakers. In an editorial, *Nature* favored the Kyoto treaty.

The *Wall Street Journal*, which did not favor Kyoto, was outraged. Its condemning editorial, "Coverup in the Greenhouse," appeared 11 June 1996. The following day, Frederick Seitz, former president of the National Academy of Sciences, detailed the illegitimate rewrite in the *Journal*, in a commentary titled, "Major Deception on Global Warming."[13]

Oddly enough, a research paper, coauthored by Santer, was published at about the same time—and says something quite different. It concludes that none of the three estimates of the natural variability of the climate spectrum agree with each other, and that until this question is resolved, "*it will be hard to say, with confidence, that an anthropogenic climate signal has or has not been detected*"[14] (emphasis added).

Why did Santer, a relatively junior scientist, make the unsupported revisions? We still don't know who directed him to do so, and then approved the changes. But Sir John Houghton, chairman of the IPCC working group, had received a letter from the U.S. State Department dated 15 November 1995. It said:

> It is essential that the chapters not be finalized prior to the completion of the discussions at the IPCC Working Group I plenary in Madrid, and that chapter authors be prevailed upon to modify their text in an appropriate manner following the discussion in Madrid.

The letter was signed by a senior career Foreign Service officer, Day Olin Mount, who was then-Acting Deputy Assistant Secretary of State. However, the Undersecretary of State for Global Affairs at that time was former Senator Timothy Wirth (D-CO). Wirth was not only an ardent advocate of man-made warming, but was a close political ally of then-President Bill Clinton and then-Vice President Al Gore. There seems little doubt that the letter was sent by Mount at the behest of Undersecretary Wirth.

Mount was later named Ambassador to Iceland. That's a plum post in a pleasant, peaceful First World country. That ambassadorship has often gone to a political ally of the White House rather than to a career diplomat.

The Madrid Plenary, held in November 1995, was a political meeting. There were representatives of ninety-six nations and fourteen nongovernment organizations (NGOs). They went over the text of the "accepted" report line by line.

Chapter 8, which should have governed the entire IPCC report, was rewritten to accord with the global warming campaign being waged by the United Nations, the NGOs, and the Clinton administration.

THE "SCIENTIFIC CONSENSUS" MANTRA

It is sheer fantasy to suggest that a huge majority of scientists with expertise in global climate change endorse an alarming interpretation of the recent climate data.

In fact, the footnoted studies in this book include hundreds of climate-science authors whose work argues against the alarmist view of climate change.

However, the assertion that there is a "scientific consensus" has played marvelously well in the newspapers and TV programs that have built their circulation with scary stories about climate change. Indications of dissent in the science community include:

- In 1992, a "Statement of Atmospheric Scientists on Greenhouse Warming" that opposed global controls on greenhouse gas emissions drew about one hundred signatures, mostly from members of technical committees of the American Meteorological Society.[15]
- The 1992 "Heidelberg Appeal," which also expressed skepticism on the urgency of restraining greenhouse gas emissions, drew more then 4,000 signatures from scientists worldwide.[16]
- The 1996 "Leipzig Declaration on Global Climate Change" emerged from an international conference on the greenhouse gas controversy, and was signed by more than one hundred scientists in climate and related fields.[17]
- A survey of more than four hundred German, American, and Canadian climate researchers was reported in the *UN Climate Change Bulletin* in 1996. Only 10 percent of the researchers surveyed said they "strongly agree" with

the statement: "We can say for certain that global warming is a process already underway." Close to half of the researchers surveyed—48 percent—said they didn't have faith in the forecasts of the global climate models, the only strong argument in favor of quick, decisive international action to counter a dangerous global warming.[18]

- A 1997 survey of U.S. State Climatologists (the official climate monitors in each of the fifty states) found 90 percent agreeing that "scientific evidence indicates variations in global temperature are likely to be naturally occurring and cyclical over very long periods of time."[19]
- In 1998, more than 17,000 scientists signed the "Oregon petition," expressing doubt about man-made global warming and opposing the Kyoto Protocol. *More than 2,600 of the signers of this anti-Kyoto petition have climate science credentials.* The petition was hosted by the Oregon Institute for Science and Medicine.[20]

The American Association of State Climatologists, in fact, has issued a contrary statement of its own:

> Climate prediction is complex, with many uncertainties. The AASC recognizes climate prediction is an extremely difficult undertaking. For time scales of a decade or more, understanding the empirical accuracy of such prediction—called "verification"—is simply impossible, since we have to wait a decade or more to assess the accuracy of the forecasts.[21]

Man-made warming proponents have tried to discredit all opposition, of course. In the case of the Oregon Petition, the detractors "discovered" a few fake names on the list of signers and reported them.

On the other hand, a group called Ozone Action, based in Washington, D.C., sent President Clinton a "Scientists' Statement on Global Climatic Disruption" in 1997. It claimed to have the signatures of 2,611 scientists from the United States and abroad, endorsing the evidence of man-made global warming as "conclusive." According to Citizens for a Sound Economy (a group opposing climate alarmism) only about 10 percent of the letter's signers had experience in fields connected with climate science. The signers did, in fact, include two landscape architects, ten psychologists, one traditionally trained Chinese doctor, and a gynecologist.[22]

THE IPCC'S "HOCKEY STICK" WIPES OUT GLOBAL CLIMATE HISTORY

Most recently, the IPCC soiled its science credentials by attempting to rewrite the Earth's known climate history. The Medieval and Roman warmings, with their intervening cold periods, present a huge problem for the advocates of man-made global warming. If the Medieval and Roman warmings occurred warmer than today—without greenhouse gases, what would be so unusual about modern times being warm as well?

Awkward World History Explained Away:

"Climate in Medieval time is often said to have been as warm as, or warmer than, it is today. Such a statement might seem innocuous. But for those opposed to action on global warming, it has become a cause celebre. If it was warmer in Medieval time than it is today, it could not have been due to fossil fuel consumption. This (so the argument goes) would demonstrate that warming in the 20th century may have been just another natural fluctuation that does not warrant political action to curb fossil fuel use . . . many new paleo-temperature series have been produced. However, well-calibrated data sets with decadal or higher resolution are still only available for a few dozen locations. . . . Only a few of these records are from the tropics, and only a handful from the Southern Hemisphere. . . . The balance of evidence does not point to a High Medieval period that was as warm as or warmer than the late 20th century." (Raymond S. Bradley, "Climate in Medieval Time," *Science* 302 (17 October 2003): 404–05.[23]

But Science Makes It More Difficult:

"A new report we release today by Dr. Diane Douglas Dalziel of Arizona State University draws upon a host of studies from sites scattered around the world that prove the existence of the Little Ice Age as a global condition not limited to the Northern Hemisphere and Europe. . . . A 41-page annotated bibliography released as part of her report includes research concerning marine cores, sea-level curves, tree-ring chronologies, peat bogs, salt marshes, stalagmites, historic records, and even human tooth enamel to determine the magnitude, timing and geographic extent of the Little Ice Age."[24]

The whole concept of global warming, natural or man-made, depends on the Earth's climate being tied into one biosphere by the air and water currents and the feedback loops. If the global climate were not a seamless, interrelated whole, people worried about "global warming" could simply move to a less industrialized part of the planet.

In 1996, as a matter of fact, the IPCC published its Second Assessment, including a graph of the past 1,000 years of global climate history showing the historical picture of Earth's recent climate variability. It showed a medieval

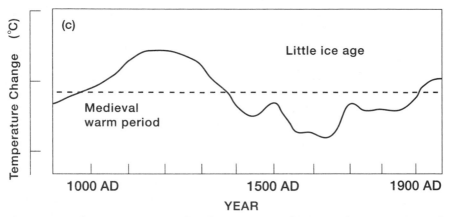

Figure 5.1.   The Last 1,000 Years of Earth Temperatures from Tree Rings, Ice Cores, and Thermometers.
(Figure 22 in the IPCC's *Climate Change,* 1995.)

warm period with warmer temperatures than today and a Little Ice Age with temperatures lower than today. (See figure 5.1.)

Six years later, the IPCC was either bolder or more desperate. In *Climate Change 2001,* the panel presented a radically different picture of the Earth's last 1,000 climate years.

Climate Change 2001 prominently displayed a graph based on a 1998 study, led by Michael Mann, a young Ph.D. from the University of Massachusetts. The Mann et al. study used several temperature proxies (but primarily tree rings) as a basis for assessing past temperature changes over the 1,000 years from 1000 to 1980. He then crudely grafted the surface temperature record of the twentieth century (much of it derived from inflated temperatures recorded by the official thermometers in urban heat islands) onto the pre-1980 proxy record.

The effect was visually dramatic. Gone were the difficult-to-explain Medieval Warming and the awkward Little Ice Age. Mann gave us nine hundred years of stable global temperatures—until about 1910. Then the twentieth century's temperatures seem to rocket upward out of control. The Mann graph became infamous in scientific circles as "the hockey stick," a shape that it resembled.[25] (See figure 5.2.)

In the United States, the Clinton administration also picked up the Mann graph and featured it as the first visual in the *U.S. National Assessment of the Potential Consequences of Climate Variability and Change*, published in 2000.

The Mann study contradicted hundreds of historical sources on the Medieval Warming and Little Ice Age, and hundreds of previous scientific pa-

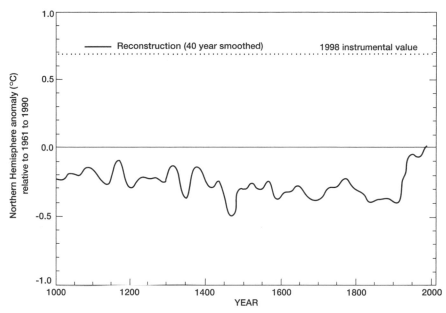

Figure 5.2:   The Infamous "Hockey Stick" Graph that Eliminates the Little Ice Age and Medieval Warming, and Exaggerates the Modern Warming.
(Figure 2.20 in the IPCC's *Climate Change* 2001.)

pers with evidence of those major past changes in the Earth's climate. But the Mann study gave the IPCC and the Clinton administration the quick answer they wanted to the historic variability of global temperatures: it never happened.

Mann was named an IPCC lead author and an editor of *The Journal of Climate*, a major professional journal, signaling the new order of things to the rest of his profession.

To affirm the apparent elimination of the awkward world history snag, the following language was included in the IPCC's *Climate Change 2001*:

> *2.3.3 Was There a "Little Ice Age" and a "Medieval Warm Period"?* The terms "Little Ice Age" and "Medieval Warm Period" have been used to describe two past climate epochs in Europe and neighboring regions during roughly the 17th to 19th and 11th to 14th centuries, respectively. . . . [V]iewed hemispherically, the "Little Ice Age can only be considered as a modest cooling of the Northern Hemisphere during this period of less than 1 degree C relative to late 20th century levels. . . . The "Little Ice Age" appears to have been most clearly expressed in the North Atlantic region as altered patterns of atmospheric circulation. . . . The evidence for temperature changes in past centuries

in the Southern Hemisphere is quite sparse. What evidence is available at the hemispheric scale for summer and annual mean conditions suggests markedly different behaviour from the Northern Hemisphere. The only obvious similarity is the unprecedented warmth of the late 20th century.[26]

In effect, the IPCC told us to ignore the overwhelming historic and physical evidence that the world's mean temperatures dropped sharply (2 to 4 degrees Fahrenheit) from about 1300 until at least 1850. They pretended this well-documented spasm of freezing cold, advancing glaciers, powerful contrary winds and terrible storms didn't freeze the Norse settlers to death on Greenland. They suggested that Earth's climate didn't get colder during the Little Ice Age. Global circulation just got a little "constricted."

If the world's climatic circulation can fail to function for five hundred years at a stretch, however, there is no point in worrying about global warming. Only your local thermometer would matter.

Fortunately, of course, there is nothing in the Earth's 3,000 years of oral or written history to support such a concept; nor is there anything in the ice cores, tree rings, or other physical evidence from paleontologists. Temperatures around the world have continued to be related to the total heat received from the sun and its complex diffusion by air and water currents.

The Damning Flaw in the "Hockey Stick"

The Mann et al. studies were controversial, despite their wide circulation. Oddly, however, it was not peer review by climate scientists that forced the Mann team to publish a "correction of error" in *Nature* on 1 July 2004.[27] Instead, it was the work of two Canadian nonscientists who were well trained in statistics—metals expert Stephen McIntyre of Toronto and economist Ross McKitrick from Canada's University of Guelph.

After the Mann studies were published in the IPCC report, McIntyre and McKitrick requested the original study data from Mann. It was provided—haltingly and incompletely—indicating that no one else had previously requested the data for a peer review in connection with the original publication in *Nature*! They found that the data did not produce the claimed results "due to collation errors, unjustifiable truncation or extrapolation of source data, obsolete data, geographical location errors, incorrect calculation of principal components and other quality control defects." Using corrected and updated source data, the two recalculated Northern Hemisphere temperature index for the period 1400–1980 using Mann's own methodology. This was published in *Energy & Environment,* with the data refereed by the World Data Center for Paleoclimatology.[28]

"The major finding is that the [warming] in the early 15th century exceed[s] any [warming] in the 20th century," report McIntyre and McKitrick. In other words, the Mann study was fundamentally wrong. Mann and his team do not yet admit this. Their "correction" specifically says that while their published proxy data set contained several errors, "None of these errors affect our previously published results."[29]

Terence Corcoran's commentary on this Mann embarrassment appeared in Canada's *Financial Post*:

> One of the great propaganda icons of the United Nations climate change machine, and the Kyoto process, is about to get swept away as a piece of junk science. The icon is The Hockey Stick, a nifty graphic that claimed to show that the world climate drifted along at nice, stable temperatures for almost 1,000 years until the late 20th century, when temperatures suddenly started to soar. News that the Hockey Stick, reproduced and cited in thousands of reports and publications, is about to get zapped is sweeping the climate science community. . . . This should come as a major embarrassment at the Intergovernmental Panel on Climate Change, the UN agency that has been using the Hockey Stick as a central propaganda tool. . . . Other scientists have also deconstructed parts of the Stick and found it to be unsupportable. And that means one of the great climate claims, that 20th century carbon emissions caused unprecedented global warming, is just plain wrong.[30]

It may be worse than wrong.

The Mann Research Left Out CO_2 Fertilization

In their exchanges with the Mann research team, McIntyre and McKitrick learned that the Mann studies give by far the heaviest twentieth-century weight in their study to the tree ring data from fourteen sites in California's Sierra Nevada Mountains. At those sites, ancient, slow-growing, high-elevation bristlecone pine trees (which can live 5,000 years) showed a strong growth spurt after 1900.

Here, finally, is the true depth of the Mann team's misrepresentation. The growth ring data from those trees were collected and presented in a 1993 paper by Donald Graybill and Sherwood Idso. It was titled: "Detecting the Aerial Fertilization Effect of Atmospheric CO_2 Enrichment in Tree Ring Chronologies."[31] Graybill and Idso specifically pointed out in their study that neither local nor regional temperature changes could account for the twentieth century growth spurt in those already-mature trees.

CO_2 acts like fertilizer for trees and plants and also increases their water use efficiency. All trees with more CO_2 in their atmosphere are very likely to grow more rapidly. Trees like the high-altitude bristlecone pines, on the margins of both moisture and fertility, are likely to exhibit very strong responses to CO_2 enrichment—which was the point of the Graybill and Idso study.

Mann and his coauthors could hardly have escaped knowing the CO_2 reality, since it was clearly presented *in the title of the study from which they derived their most heavily weighted data sites*.

THE BIGGEST GLOBAL WARMING FALSEHOOD:
IMPLEMENTING KYOTO—CHEAP AND EASY

The biggest fraud associated with the Kyoto Protocol is the idea that it will be cheap and easy to shift humanity from fossil fuels to "renewables" such as wind and solar power. Kyoto supporters usually maintain that it's just the big oil companies and the electric power monopoly that have captured our minds and kept us in thrall to fossil fuels for all these years. Few Kyoto advocates suggest using nuclear power, the only current cost-effective energy alternative that produces no CO_2.

Kyoto plays into this renewable mind-set by pretending that its "5 percent solution" would curb greenhouse gases. In reality, as the IPCC itself has indicated, stabilizing atmospheric greenhouse gas concentrations at "moderate" levels might require cutting fossil fuel use by 60 to 80 percent *worldwide*. Kyoto member countries might have to give up virtually all of their fossil fuels to achieve that questionable goal, given the rising emissions of developing countries.

Drastically reducing global atmospheric CO_2 levels through the Kyoto mechanism would clearly force a radical change in humanity's use of energy and technology. That's what it's supposed to do.

The world's people would have to invest in a controversial shift to nuclear power or almost completely renounce modern technology. Neither course of action is yet acceptable to very many of the world's citizens.

Given the urgent desire of the Third World to live as well as Americans and Europeans, it seems even less likely that Third World countries will sign Kyoto than that the First World countries will rapidly shift to renewable energy sources.

Kyoto members would have to politically ration the lighting, heating, and refrigeration for their homes, schools, hospitals, factories, and businesses. Transportation for manufactured goods and off-season fresh fruits and vegetables would be sharply curtailed. Tourism might be possible only for the winners of a "travel lottery." Standards of living would plummet, with far fewer good jobs, fewer attractive lifestyle choices, and fewer ways to improve human health.

Much of the world's landscape would be covered with solar panels and wind turbines. (The electricity from all of the thousands of acres of windmills and solar panels in California today equal the output from just one good-sized electrical generating plant using twenty acres.) Nor do solar, wind, or even nuclear energy produce a liquid fuel for cars, trucks, and planes.

Billions of people in the Third World would find it vastly harder to rise out of poverty, and move away from the unsustainable use of supposedly renewable wood, straw, and dung for cooking and heating. (The straw and dung need to go back on the fields to maintain the fields' soil health.)

The fate of wildlife would be heavily affected as well, with much of the world's remaining forests and wildlife habitat at risk of being cleared for firewood and low-yield crops.

NOTES

1. "About Climate Action Network Europe," <www.climnet.org/aboutcne.htm>.

2. "About USCAN," 30 January 2006, <www.usclimatenetwork.org/about>.

3. Ken Green, "The 'Fatal Conceit' of Kyoto," *Toronto Star*, 25 April 2004.

4. Quoted by the Associated Press, 6 April 1990.

5. The Byrd-Hagel Resolution passed 95–0 by the U.S. Senate on 21 July 1997.

6. Byrd-Hagel Resolution.

7. M. Parry et al., "Adapting to the Inevitable," *Nature* 395 (1998): 741.

8. Intergovernmental Panel on Climate Change, *First Assessment Report* (Cambridge, UK: Cambridge University Press, 1990).

9. IPCC, *Second Assessment Report*, 1996, Summary for Policymakers.

10. IPCC, *Third Assessment Report*, 2001.

11. IPCC, *Second Assessment Report*, 1996, chapter 8, 412.

12. IPCC, *Second Assessment Report*, 1996, chapter 8, 439.

13. F. Seitz, "Major Deception on Global Warming," *Wall Street Journal*, 12 June 1996.

14. T. P. Barnett, B. D. Santer, P. D. Jones, R. S. Bradley, and I. R. Briffa, "Estimates of Low-Frequency Natural Variability in Near-Surface Air Temperatures," *The Holocene* 6 (1996): 96.

15. S. Fred Singer, *Hot Talk, Cold Science* (Washington, D.C.; Independent Press, 1992), 40. Quoted statement by Atmospheric Scientists on Greenhouse Warming, 27 February 1996.

16. Heidelberg Appeal, publicly released at the 1992 Earth Summit in Rio de Janeiro, www.sepp.org/heidelberg_appeal.html

17. Text of the Leipzig Declaration on Global Climate Change, <www.sepp.org/leipzig.html>.

18. Dennis Bray and Hans von Storch, "1996 Survey of Climate Scientist on Attitudes towards Global Warming and Related Matters," *Bulletin of the American Meteorological Society* 80 (March 1999): 439–55.

19. "Survey of State Experts Casts Doubt on Link between Human Activity and Global Warming," press release, 1997, Citizens for a Sound Economy, Washington, D.C.

20. Global Warming Petition, Oregon Institute for Science and Medicine, Cave Junction, Oregon, <www.oism.org/pproject/>.

21. *Policy Statement on Climate Variability and Change*, American Association of State Climatologists, approved November 2001.

22. Scientists' Statement on Global Climatic Disruption, Ozone Action, Washington, D.C., 6 June 1997.

23. Raymond S. Bradley, "Climate in Medieval Time," *Science* 302 (17 October 2003): 404–05. Bradley coauthored with Michael Mann two widely cited studies denying that the Medieval Warming and Little Ice Age were important global climate events.

24. Fred Palmer, President, Greening Earth Society, 25 January 2001.

25. M. E. Mann et al., "Global-Scale Temperature Patterns and Climate Forcing over the Past Six Centuries," *Nature* 392 (1998): 779–87.

26. IPCC, *Climate Change 2001*: The Scientific Basis: Chapter 2, Section 2.3.3, "Observed Variability and Change," www.grida.no/climate/ipcc_tar/wgl/070.htm.

27. Mann et al., "Corrigendum: Global-Scale Temperature Patterns and Climate Forcing over the Past Six Centuries," *Nature* 430 (1 July 2004): 105.

28. S. McIntyre and R. McKitrick, "Corrections to the Mann et al., "Proxy Data Base and Northern Hemispheric Average Temperature Series, 1998," *Energy & Environment* 14 (2003): 751–71.

29. Mann et al., "Corrigendum," 105.

30. T. Corcoran, "The Broken Stick," *Financial Post*, 13 July 2004.

31. D. A. Graybill and S. B. Idso, "Detecting the Aerial Fertilization Effect of Atmospheric CO_2 Enrichment in Tree Ring Chronologies," *Global Biogeochemical Cycles* 7 (1993): 81–95.

Chapter Six

The Baseless Fears

A Million Wild Species Will Be Lost Forever

Extinction Believers Cry:

"The midrange estimate is that 24 percent of plants and animals will be committed to extinction by 2050," said ecologist Chris Thomas of Britain's University of Leeds. "We're not talking about the occasional extinction—we're talking about 1.25 million species. It's a massive number."[1]

"Many biologists, myself included, believe we're standing at the edge of a mass extinction," said Terry Root, a senior fellow with the Institute for International Studies at Stanford University. "A lot of species aren't going to make it because they have to change and move to new locations so rapidly. And how can they do that when their habitat has been destroyed?"[2]

Reality-Based Skeptics See Evolution Differently:

"Scientists said yesterday they have found evidence that a huge meteorite or comet plunged into the coastal waters of the Southern Hemisphere 251 million years ago, possibly triggering the most catastrophic mass extinction in Earth's history. The researchers said that geological evidence suggests that an object about six miles in diameter crashed at the shoreline of what is now Australia's northwestern coast, creating climate changes and other natural catastrophes that wiped out 90 percent of marine species and 70 percent of land species."[3]

Man-made global warming could destroy more than one million plant and animal species in the next fifty years, warned an international scientific research team through press releases and in *Nature* in early 2004. The nineteen-member team was led by Chris D. Thomas of Britain's University of Leeds.

The Thomas report said global climate models predict such radical changes in ecosystems and eco-processes that they will shift species' "survival envelopes" and make it impossible for huge numbers of birds, animals, fish, and lower life forms to continue living on the planet.[4] The Thomas report received huge amounts of coverage in the press. This report followed two earlier, equally gloomy extinction projections in *Nature* from smaller U.S. teams, one led by Terry Root of Stanford University, and the other by Camille Parmesan of the University of Texas.[5] All of these researchers concluded that plants and animals will be unable to keep up with the unprecedented speed and scale of temperature change in the next fifty years and species will go extinct in vast numbers.

The rationale for the extinctions put forth by the Thomas team had been discredited, however, before their paper was published. Some of the contrary evidence was published by Chris D. Thomas himself.

FAILING THE COMMON-SENSE TEST

Most of the world's animal species "body types" were laid down during the Cambrian period, 600 million years ago, according to Jeffrey Levinton, chairman of the Department of Evolution and Ecology at the State University of New York–Stony Brook in a widely noted 1992 article in *Scientific American*.[6] Thus, we know that the major species have dealt successfully through the ages with new pest enemies, new diseases, ice ages, and global warmings higher than today's.

Virtually every wild species is at least one million years old, which means that they've all been through at least six hundred of the 1,500-year climate cycles. Not the least of the warmings was the Holocene Climate Optimum, which was warmer than even the predictions of the Intergovernmental Panel on Climate Change for 2100. That very warm period lasted for 4,000 years and ended less than 5,000 years ago.

Environmentalists argue that the speed of today's climate change is much greater than ever before and it will overwhelm the adaptive capacities of the critters and conifers. Yet history and paleontology agree that many of the past global temperature changes arrived very quickly, sometimes in a few decades. As an example, 12,000 years ago the Younger Dryas event, suddenly and violently, swung from warm temperatures back to ice age levels.

The warming was then thrown into reverse by the shutdown of the Gulf Stream, as melting water from the extra trillions of tons of ice built up in the glaciers and ice sheets over the previous 90,000 years of frigid climate was released into the oceans. The shutdown of the oceans' Atlantic Conveyor quickly triggered another thousand years of Ice Age. How did wild species deal with Mother Nature's sudden, sharp reversals then?

In another example, in the 1840s, a Wyoming glacier went from Little Ice Age cold to near Modern Warming warmth in about a decade.[7] But there's no evidence of species going extinct from the rapid temperature change.

THE PROVEN EXTINCTION THREATS: ASTEROIDS, HUNTING, FARMING, AND TRAVEL

We already know how most of the world's extinct species were lost, and in what order of magnitude.

First: Huge asteroids impacting the Earth. The Web sites of such universities as the University of California–Berkeley, Smith, and North Carolina State are replete with evidence of these "big bangs." The Earth's collisions with massive missiles from outer space explode billions of tons of ash and debris into the Earth's atmosphere, darken the skies, and virtually eradicate the growing seasons for years at a time. There have apparently been more than a dozen such massive collisions in the Earth's past, and they have destroyed millions of species, most of which we know about only through the fossil record.

Researchers recently announced that geological evidence which suggests an object about six miles in diameter crashed at the shoreline of what is now Australia's northwestern coast 251 million years ago, creating climate changes and other natural catastrophes that wiped out 90 percent of marine species and 70 percent of land species.[8]

Second: Hunting. The next most massive assault on our wild species has come from humanity's attempts to feed itself. For a million years or so, we hunted whatever we could catch. If it went extinct, we hunted something else. In this sense, the last ice age did cause some indirect species extinctions. During that extremely cold period, so much of the world's water was trapped in ice caps and glaciers that sea levels dropped about four hundred feet below today's levels. Stone Age hunters walked across the Bering Strait from Asia and found hordes of wild birds and mammals that did not fear man. More than forty edible species were wiped out in a historical eyeblink, including North America's mammoths, mastodons, horses, camels, and ground sloths.[9]

Third: Man learned to farm. Farming for food made us less likely to hunt wild animals and birds to extinction, but eventually we claimed one-third of all the Earth's land area for agriculture. The saving grace was that the best land for farming tended to have few species; instead it had large numbers of a few species, such as bison on the American Great Plains and the kangaroo in the Australian grasslands. In contrast, researchers have found as many species in five square miles of the Amazon as in the whole of North America. Fortunately, man tended to farm the best land for the highest sustainable food yields, leaving much of the poorer land (with its diversity) for nature.[10]

Fourth: Alien species. Mankind's ships, cars, and planes all transport alien species, sometimes knowingly and more often unwittingly. This has made the survival competition among species much more global. Island species, in particular, have found themselves in more intense competition and many have gone extinct.

THE "SCIENTIFIC" EVIDENCE OF EXTINCTIONS FROM THE MODERN WARMING

The Thomas team said that simply raising or lowering the Earth's temperature would cause major wildlife extinctions on a linear model. Small temperature changes would lead to relatively small reductions in species numbers, and larger temperature increases would drive far more extinctions.

The team first defined "survival envelopes" for more than 1,100 wildlife species—in Europe, the Brazilian Amazon, the wet tropics of northeastern Australia, the Mexican desert, and the southern tip of South Africa. Then they used a "power equation" to link loss of habitat area with extinction rates. If the equation showed that a species' potential habitat was projected to decline, it was regarded as threatened; the greater the expected habitat loss, the greater the threat. There was no provision for species adaptation or migration.

One of the Thomas team's "moderate" scenarios was an increase in Earth's temperature of 0.8 degrees Celsius in the next fifty years. The researchers said this would cause the extinction of roughly 20 percent of the world's wild species, perhaps one million of them.

Fortunately, this prediction can easily be checked. The Earth's temperature has already increased at least 0.6 degrees Celsius over the past 150 years. How many species died out because of that temperature increase?

The answer? None.

The Thomas paper tells us in its opening sentence: "Climate change over the past 30 years has produced numerous shifts in the distributions and abundances of species, and has been implicated in one species-level extinction."[11]

That's right. The scientists who are predicting that 0.8 degrees Celsius of warming would cause hundreds of thousands of wildlife species extinctions over the next 50 years concede that roughly this level of temperature increase over the past 150 years has resulted in the extinction of *one* species.

Reality takes away even that one extinction claim.

Thomas's single cited example of a species driven extinct by the recent warming is the Golden Toad of Costa Rica. That was based on a 1999 paper in *Nature* by J. Alan Pounds, at the Monteverde Cloud Forest Preserve in Puntarenas, Costa Rica.[12] Pounds claimed that, due to rising sea surface temperatures in the equatorial Pacific, twenty of the fifty species of frogs and toads (including the Golden Toad) had disappeared in a cloud forest study area of 30 square kilometers. (Cloud forests are misty habitats found only in the mountains above 1,500 meters, where the trees are enclosed by cool, wet clouds much of the time. The unusual climate serves as a home to thousands of unique plants and animals.)

Pounds explained his thesis to a scientific conference in 1999:

> In a cloud forest, moisture is ordinarily plentiful. Even during the dry season . . . clouds and mist normally keep the forest wet. Trade winds, blowing in from the Caribbean, carry moisture up the mountain slopes, where it condenses to form a large cloud deck that surrounds the mountains. It is hypothesized that climate warming, particularly since the mid-1970s, has raised the average altitude at which cloud formation begins, thereby reducing the clouds' effectiveness in delivering moisture to the forest. . . [D]ays *without* mist during the dry season. . . . quadrupled over recent decades.[13]

Pounds says at least twenty-two species of amphibians have disappeared from the cloud forest. Although the other species that disappeared were known to exist in other locations, the Golden Toad lost its only known home.

However, two years after Pounds hypothesized that the amphibians lost their cloud forest climate to drying from sea surface warming, another research team demonstrated that it was almost certainly the clearing of lowland forests under the cloud forest of Monteverde that changed the pattern of cloud formation over the Golden Toad's once-mistier home.[14]

The team, led by R. O. Lawton from the University of Alabama–Huntsville, noted that trade winds bringing moisture from the Caribbean spend five to ten hours over the lowlands before they reach the Golden Toad's mountain home in the Cordillera de Tilerán. By 1992, only about 18 percent of the original lowland vegetation remained. The deforestation reduced the infiltration of rainfall, increased water runoff, and thus reduced soil moisture. The shift from trees to crops and pasture also reduced the amount of water-holding canopy.

In March 1999, the Lawton team got satellite imagery showing that late-morning dry season cumulus clouds were much less abundant over the deforested parts of Costa Rica than over the nearby still-forested lowlands of Nicaragua.

To check their conclusion, the Lawton team simulated the impact of Costa Rican deforestation on Colorado State University's Regional Atmospheric Modeling System. The computer modeling showed that the cloud base over pastured landscape rose above the altitude of the Cordellera peaks (1,800 meters) by late morning. Over forests, the cloud base didn't reach 1,800 meters until early afternoon. Lawton says these values "are in reasonable agreement with observed cloud bases in the area."[15]

That puts the blame for the cloud forest dryness squarely on the farmers and ranchers who cleared the lowlands. *Pounds' own paper noted deforestation as a major threat to mountain cloud forests.* The Lawton study leaves the Thomas team's big computerized study of mass species extinction without any evidence that moderate climate changes—even when abrupt—cause species extinctions.

Didn't anybody on the team question how they could justify predicting one million extinctions from 0.8 degrees of warming when the only evidence for their thesis was a shaky hypothesis based on a few toads and lizards in one cloud forest? Didn't they at least do an Internet search to ensure that their Golden Toad extinction theory would hold up under scientific review?

HIGH-YIELD FARMING REDUCES WILDLIFE EXTINCTIONS

In 2002, the United Nations Environmental Program (UNEP) published a new *World Atlas of Biodiversity.* They reported that the world lost only half as many major wild species in the last three decades of the twentieth century (twenty birds, mammals, and fish) as during the last three decades of the nineteenth century (forty extinctions of major species). In fact, UNEP said the rate of extinctions at the end of the twentieth century was the lowest since the sixteenth century—despite 150 years of rapidly rising world temperatures.

The big reason for the reduction in species losses had nothing to do with temperatures. It had to do with improved agriculture. Higher-yield seed varieties, irrigation, chemical fertilizer, and pesticides tripled the crop yields on most of the world's good farmland after 1960. Humanity was no longer forced to clear large tracts of additional land for low-yield crops.

The Swiss-based World Conservation Union (IUCN) is arguably the largest and most professional wildlife conservation group in the world. In 2001, the IUCN published a report titled *Common Ground, Common Future.* The report's authors stated that the largest current threat to wildlife species in the world is the billion poor people trying to subsist in the world's "hot spots" of biodiversity (most of them tropical). These people currently hunt bush-meat with such new technologies as AK-47s and commit low-input, slash-and-burn farming to support the large families that subsistence societies almost always have.[16]

The IUCN recommended "eco-agriculture" as the solution. Essentially, this means showing low-yield farmers how to get more food per acre so that they can leave more of their poorer-quality land as habitat, refuges, and corridors for wildlife. In effect, they are endorsing the argument of Norman Borlaug, the 1970 Nobel Peace Prize laureate. Borlaug said: "Growing more food per acre leaves more land for Nature."[17]

Such high-yield conservation could have forestalled the need to deforest the Costa Rican lowlands that evidently doomed the Golden Toad.

Ironically, many of the same eco-groups claiming man-made warming will destroy vast numbers of species have also been campaigning against agriculture's Green Revolution and nitrogen fertilizer, both of which have forestalled much greater deforestation. Worse, the Kyoto treaty is likely to make modern technology (including nitrogen fertilizer) so energy expensive that economic growth and agricultural improvements will be virtually impossible in the Third World.

Then, the billion subsistence farmers in the key biodiversity regions would needlessly threaten many more species than the Golden Toad.

THE ECOLOGICAL CHANGES DUE TO GLOBAL WARMING WON'T BE RAPID

The alarmists' key argument for massive extinctions is that trees and plants are essentially immobile. They won't be able to move rapidly, even if rapid global warming pushes them beyond their "survival envelopes." Then the mammals and lower orders of creatures will go extinct because the trees and plants on which so many of them depend will have disappeared. Rapid warming will irreparably shred the vital "web of life."

Not exactly.

The trees and plants won't have to change rapidly. Trees and plants that thrive nearer the poles have adapted to cold, *but their range is rarely limited by warmth.*

In the Northern Hemisphere, for example, Craig Loehle of the Argonne National Laboratory notes that cold-adapted trees can grow to maturity only fifty to one hundred miles north of their natural ranges.[18] However, they can often grow to maturity as much as one thousand miles south of their southern boundaries. Loehle reports that "many alpine and arctic plants are extremely tolerant of high temperatures, and in general one cannot distinguish between arctic, temperate, and tropical moist-habitat types on the basis of heat tolerances."[19] This means, according to Loehle, "Seedlings of these southern species will not gain much competitive advantage from faster growth in the face of existing stands of northern species. The existing adult trees have such an advantage due to light interception. Southern types must wait for gap replacement, disturbances, or stand breakup to utilize their faster growth to gain a position in the stand. This ensures that the replacement of species will be delayed at least until the existing trees die, which can be hundreds of years. . . . Thus, the replacement of forest (southern types replacing northern types) will be an inherently slow process (between several and many hundreds of years)."[20]

This means that the mammals, birds, fish, lichens, mushrooms, and other species that depend on the plant life and plant-based ecosystems will have ample time to shift with them.

The Specter of Species Extinction, a recent study published by the Marshall Institute, is even more optimistic than Loehle and just as powerfully condemns the thesis that moderate warming will drive wild species to extinction. The authors of this report argue that *global warming will bring more species diversity, not less, to most parts of the globe*. Rather than wiping out species, the Marshall study predicts that global temperatures will extend the ranges of thousands of plants and animals, and encourage forests and meadows with a greater diversity of plants, trees, and dependent species.[21]

The Specter of Species Extinction was written by a father-and-sons research team, led by climate physicist, Sherwood B. Idso, and his sons, Craig (a specialist in climate geography) and Keith E. (a botanist who specializes in plants' response to CO_2 changes). The Idsos' report stresses that warmer temperatures give most trees, plants, animals, and fish the opportunity to extend their ranges toward the North and South Poles without imposing any "heat limits" that would force them to give up the ranges they currently occupy. They agree with Loehle's conclusion that only over hundreds of years would the faster growing trees from the south be able to out-compete the already mature northern trees. Forests and plants would shift their ranges northward and southward very slowly.

Another possibility that must be seriously considered, the authors say, is that northern or high-altitude forests will not be replaced at all by southern or low-altitude forests in a warming world. Rather, the two forest types may merge, creating entirely new forests of greater species diversity. The research of D. I. Axelrod has found that such species-rich forests existed during the warmer Tertiary Period of the Cenozoic Era, when in the western United States many (mountain species) regularly grew among mixed conifers and broadleaf schlerophylls.[22]

PLANTS MIGHT NOT HAVE TO SHIFT AT ALL, THANKS TO HIGHER CO2 LEVELS

The Idso analysis also notes that higher CO_2 levels act as fertilizer for trees and plants, and that higher CO_2 levels also reduce the amount of energy needed by most plant species to conduct a process called photorespiration. *So long as temperatures and CO_2 are both rising, trees and plants gain vigor* with which to exploit warming's opportunities for range expansion.

"Proponents of what we shall call the CO_2-induced global warming extinction hypothesis seem to be totally unaware of the fact that atmospheric CO_2 enrichment tends to ameliorate the deleterious effects of rising temperatures on Earth's vegetation," says the Idso report. "They appear not to know that more CO_2 in the air enables plants to grow better at nearly all temperatures, but especially at higher temperatures. . . . Under such conditions, plants living near the heat-limited boundaries of their ranges do not experience an impetus to migrate poleward or upward towards cooler regions of the globe. At the other end of the temperature spectrum, however, plants living near the cold-limited boundaries of their ranges are empowered to extend their ranges into areas where the temperature was previously too low for them to survive. . . .

"Animals react in much the same way. Over the past century and a half of increasing air temperature and CO_2 concentration, many species of animals have significantly extended the cold-limited boundaries of their ranges, both poleward in latitude and upward in elevation, while they have maintained the locations of the heat-limited boundaries of their ranges."[23]

The Idsos are widely published on CO_2 benefits to plant growth and the phenomenon has been widely studied in dozens of countries because of its importance to crop growth. Their peer-reviewed analysis of forty-two experimental data sets collected by numerous scientists showed that the mean growth enhancement from a 300 parts-per-million increase in atmospheric CO_2 rises from almost nil at 10°C to *doubled growth at 38°C.*[24] At higher temperatures, the growth stimulation rises even higher.[25]

The importance of CO_2 as a fertilizer is endorsed by satellite observations of global vegetation from 1982 to 1999, which found an increase in global plant growth of more than 6 percent. The planet during that period featured slightly increasing rainfall and slightly rising temperatures—but the major change for plants was the rapid increase in atmospheric CO_2.[26] All of the regions showed positive gains in plant growth—despite the real and imagined environmental stresses that climate warming alarmists have been telling us threaten the world's plant life.

Doubling the level of CO_2 raises the net productivity of herbaceous plants by 30 to 50 percent and of trees and woody plants by 50 to 80 percent, based on extensive reviews of the research by Sherwood B. Idso and Bruce A. Kimball, then of the U.S. Department of Agriculture's Water Conservation Laboratory, and Henrik Saxe, of the Royal Veterinary and Agricultural School of Denmark.[27]

Even if the planet warms significantly because of higher CO_2 levels, the vast majority of Earth's plants will apparently not feel much need to migrate toward cooler parts of the globe.

MORE MISLEADING "EXTINCTION" STUDIES

Two other recent articles in *Nature* offer scary extinction predictions. One of these studies was led by Stanford's Terry Root. Her coauthors included, among others, J. Alan Pounds of Golden Toad fame, and her husband, Stephen Schneider, one of the most prominent advocates for the theory that global warming is man-made and dangerous. Their paper, titled "Fingerprints of Global Warming on Wild Animals and Plants," was published in the 16 January 2003 issue of *Nature*.[28] The other *Nature* study, published in the same edition, was led by Camille Parmesan of the University of Texas. It was titled, "A Globally Coherent Fingerprint of Climate Change Impacts across Natural Systems."[29]

Amazingly, *none* of the studies reviewed by the Stanford and Texas-led teams documented an extinction threat. The closest thing to an extinction threat in the Root study was the expansion of red foxes into the southern former range of arctic foxes in North America and Eurasia. However, this is displacement/replacement, not extinction. Pall Hersteinsson of the University of Iceland and Dusty W. Macdonald of the University of Alaska concluded that the changes in fox ranges were driven by prey availability. The arctic foxes are found primarily in the treeless regions of the Arctic, where they feed on lemmings and voles in the summer and eat heavily from seal carcasses in the winter. The larger red foxes eat a wider range of prey and fruits, and are regarded as stronger competitors in forest and brush land. However, they are less well camouflaged for the winters in the treeless tundra than the arctic foxes in their blue-white winter pelts.[30]

Warming temperatures have allowed trees, brush, and red foxes to move farther north in the past 150 years—but they have also allowed arctic foxes to retain enough land and prey to succeed north of the red foxes. We do not know what would have happened if there had been no northern habitat and prey to support the arctic foxes during a red fox expansion, but the foxes have already survived more radical warming than they have faced recently. In earlier parts of the interglacial period, the Arctic temperatures were 2 to 6 degrees Celsius higher than they are now.[31]

CHRIS D. THOMAS'S EARLIER STUDY
CONTRADICTS HIS *NATURE* PAPER

Chris D. Thomas's headline-grabbing computer study claiming a million species will be wiped out was written after the Root and Parmesan studies, and after *The Specter of Species Extinction*. However, a study by his earlier team was included in the Root analysis. Amazingly, it completely discredits the thesis on which Thomas based his 2004 *Nature* claim—that species have readily defined "survival envelopes" outside which they cannot survive.

The Thomas team began their 2001 paper by restating the long-believed and broadly held concept that many animals are "relatively sedentary and specialized in marginal parts of their geographical distributions." Thus, creatures are "expected to be slow at colonizing new habitats."[32]

Despite this belief, however, the Thomas team cites its own and many other researchers' studies showing that, "the cool margins of many species' distributions have expanded rapidly in association with recent climate warming."[33] This mildly undercuts their thesis.

Much worse (or much better) was to come. The two butterfly species the authors studied themselves revealed something truly startling. The butterflies "*increased the variety of habitat types that they can colonize*" (emphasis added). The Thomas 2001 report further noted that the two species of bush cricket they studied showed "increased fractions of longer-winged (dispersive) individuals in recently founded populations."[34] The longer-winged crickets would be able to fly farther in search of new habitat.

As a consequence of the new adaptations, the Thomas authors report that:

> Increased habitat breadth and dispersal tendencies have resulted in about 3 to 15-fold increases in [range] expansion rates, allowing these insects to cross habitat disjunctions that would have represented major or complete barriers to dispersal before the expansions started.[35]

Obviously, the changes in the butterfly and cricket populations render Thomas' entire thesis of "survival envelopes" inadequate at best and quite likely irrelevant. Yet this paper was not only written before the Root analysis, it was included in the Root analysis as one of the select few research studies supporting the mega-extinction theory.

Other biologists have found additional evidence of the adaptability of species. Biologists in the 1990s reported that mudworms in Foundry Cove of New York's Hudson River—near an old battery factory—had developed an amazing resistance to cadmium. "The evolution of cadmium resistance could have taken no more than 30 years," said evolutionary biologist Jeffrey

Levinton of the State University of New York–Stony Brook. "This capacity for rapid evolutionary change in the face of a novel environmental challenge was startling. No population of worms in nature could ever have faced conditions like the ones humankind created in Foundry Cove. . . . The rapid revolution of tolerance for high concentrations of toxins seems to be common. Whenever a new pesticide is brought into use, a resistant strain of pest evolves, usually within a few years. The same thing happens to bacteria when new antibiotics are introduced."[36]

An Antarctic fish, the *Pagothenia borchgrevinki*, has recently been found to tolerate temperatures up to 9 degrees Celsius warmer than the stable –1.9°C. that have characterized Antarctic waters for the past 14 million years—with an annual variation of less than 0.3 degrees Celsius. Thus the fish could apparently survive even if global warming completely melted the Antarctic ice cap, says Cara Lowe, the New Zealand graduate student who reported the research to the New Zealand Antarctic Conference at Waikato University in 2004.[37]

In 2004, a group of Oxford academics published their own critique of the "mass extinction" papers in a paper titled *Crying Wolf on Climate Change and Extinction*.[38] The four authors are members of the biodiversity research group at Oxford's school of geography and the environment. Their critique accused environmental groups of exaggerating the threat of climate change to raise more money from public donations. Dr. Paul Jepson, one of the critique's authors, said, "Even the idea of modeling climate change is mostly theoretical at the moment, so it is impossible to make these kind of predictions or counter-predictions about species." Dr. Richard Ladle, another of the critical academics, said that most species cited as being in danger of extinction by 2050 probably won't be. Ladle said the response of the World Wildlife Fund to their criticism was that the environmental ends justified simplifying appeals to make them more emotional.[39]

REAL-WORLD STUDIES FIND
TREND TOWARD SPECIES RICHNESS

Instead of supporting mass extinctions, the Root and Parmesan studies broadly confirm the Idso-Loehle thesis—enhanced species richness with warming and CO_2 increase.

Parmesan herself examined the northern boundaries of fifty-two butterfly species in northern Europe and the southern boundaries of forty butterfly species in southern Europe and North Africa over the past century. Given the 0.6 degrees Celsius warming of Europe's temperatures over that period, it is striking to have Parmesan tell us that "nearly all northward shifts involved extensions at the northern boundary with the southern boundary remaining stable." Thus, "most species effectively expanded the size of their ranges."[40]

Between 1970 and 1990, Chris Thomas documented changes in the distribution of many British bird species.[41] He found that the northern margins of southerly species shifted northward by an average of nineteen kilometers, while the southern margins of northerly species remained unchanged.

On twenty-six high mountain summits in the middle part of the Alps, a study of the plant species found "species richness has increased during the past few decades, and is more pronounced at lower altitudes." In other words, the mountain tops show *little loss of biodiversity at upper elevations, and increased species richness at lower elevations, where plants from still-lower elevations extended their ranges upward.*[42]

The University of Vienna's Harald Pauli examined the summit flora on thirty mountains in the European Alps, with species counts that ranged back in history to 1895. He says mountaintop temperatures have risen by 2 degrees Celsius since 1920, with an increase of 1.2 degrees Celsius in just the last thirty years. Nine of the thirty mountaintops showed no change in species counts, but eleven gained an average of 59 percent more species, and one mountain gained an astounding 143 percent in species numbers. Did historic species get crowded out by the flood of new warmer-zone plants? The thirty mountains showed a mean species loss of 0.68 out of an average of 15.57 species.[43] (There was no documentation that any of the species "lost" on particular mountains represented extinctions rather than local disappearances.)

Native heat-sensitive plant species have responded to temperature increases in the Iberian Peninsula and the Mediterranean coast over the past thirty years *by expanding their ranges* "towards colder inland areas where they were previously absent."[44]

The number of lichen species groups present in the Central Netherlands increased from 95 in 1979 to 172 in 2001 as the region warmed. C. M. Van Herk of the University of Eindhoven found the average number of species grouped per site increased from 7.5 to 18.9.[45] Again, more warmth produced increased species richness.

Looking at the distribution of eighteen butterfly species widespread and common in the British countryside, "nearly all of the common species have increased in abundance [during the warming], more in the east of Britain than in the west," according to Emie Pollard of Britain's Institute of Terrestrial Ecology.[46]

Warm-water species of plankton rapidly responded to warming and cooling in the western English Channel, shifting latitudinally by up to 120 miles, and increasing or decreasing their numbers by two- to threefold over seventy years, found A. J. Southward of Britain's Marine Biological Association.[47]

Parmesan and Yohe's Additional Studies

In the Antarctic, Adelie penguins need pack ice to succeed, whereas the chinstrap penguins prefer ice-free waters. R. C. Smith found that the Adelie penguins in the West Antarctic Peninsula are declining because the warming on the peninsula favors the chinstraps. Meanwhile the chinstraps in the Ross Sea region are suffering, because the 97 percent of Antarctica that isn't the peninsula is getting colder.[48]

What's the surprise? Two varieties of a highly mobile species have moved to the sites that favor their respective feeding and breeding requirements—while their populations decline in the unfavorable sites. Fortunately, it would take thousands of years of increased warming to melt all the Antarctic ice and a very long period of extended cold to close all the open water around the Antarctic Peninsula. The 1,500-year climate cycle makes it almost certain that both types of penguins will continue to move and adapt, but neither will disappear.

The Antarctic's only two higher-level plant species have responded to the Antarctic Peninsula's warming by increasing their numbers at two widely separated localities.[49]

Invertebrates in a rocky intertidal site at Pacific Grove, California, can't move, but their populations change. The invertebrates were surveyed in 1931–1933 and again 1993–1996 (after a warming of 0.8 degrees Celsius) Ten of the eleven southern species increased in abundance, whereas five of seven northern species decreased.[50]

New photographs were taken to match a set of 1948–1950 photographs of the Brooks Range and the Arctic coast of Alaska. At more than half of the matched locations, researchers found "distinctive and, in some cases, dramatic increases in the height and diameter of individual shrubs . . . and expansion of shrubs into previously shrub-free areas."[51]

Again, the Root-selected studies seem mainly to demonstrate the aggressive efforts of species to expand their ranges when and where possible. (In the Arctic, do we thank the temperature changes or the higher levels of CO_2?)

Western American bird species are pioneering and expanding their ranges over vast areas and huge climatic differences as the climate warms. N. K. Johnson of the University of California–Berkeley compiled records for twenty-four bird species from *Audubon Field Notes*, *American Birds* and other sources. He found "four northern species have extended their ranges southward, three eastern species have expanded westward, fourteen southwestern or Mexican species have moved northward, one Great Basin-Colorado Plateau species has expanded radially, and two Great Basin-Rocky Mountain subspecies have expanded westward."[52]

WHAT IS "EXTINCTION"?

Looking at the research papers selected to support the theory of massive warming extinctions, we are struck by the biologists' apparent misunderstanding of extinction. Some biologists seem to believe that effective conservation means every local population of butterflies and mountain flowers must be preserved. This is obviously impossible on a planet with continual, massive climate changes (and human impacts). Some biologists try to define more and more local populations as separate species—a subterfuge. A recent article in *Science* amply illustrates the conflicted feelings and biologists' frantic desire to protect everything. In "All Downhill from Here," author Kevin Krajick laments the supposed danger to the cute little pikas (rodents, cousins to rabbits) that live on treeless mountaintops:

> As global temperatures rise, the pika's numbers are nose-diving in far-flung mountain ranges . . . researchers fear that if the heat keeps rising, many alpine plants and animals will face quick declines or extinctions . . . creatures everywhere are responding to warming, but mountain biota, like cold-loving polar species, have fewer options for coping. . . . Comprising just 3% of the vegetated terrestrial surface, these islands of tundra are Noah's ark refuges where whole ecosystems, often left over from glacial times, are now stranded amid un-crossable seas of warm lowlands.[53]

Krajick himself seems to forget the words in his own opening paragraph: pikas "are also some of the world's toughest mammals." As for his "uncrossable seas of warm lowlands," the pikas may not be able to thrive in the lowlands competition, but it does seem likely they could find enough vegetation there during their travels to tide them over until they find other, cooler mountaintops.[54]

There Is No Such Thing as a Local Extinction

Parmesan's *Nature* paper also presents a gross distortion of the term "extinction." Parmesan counted Edith's checkerspot butterflies at 115 North American sites with historical records of harboring the species *Euphydryas editha*.

She classified the sites as "extinct or intact." She found the local checkerspot populations in much-warmer Mexico were four times more likely to be "locally extinct" than those in much-cooler Canada.

However, there is no such term as "locally extinct." Extinct means "no longer in existence; died out." Gone forever. Parmesan is using it here in reference to butterfly populations that have simply moved—and even left forwarding addresses farther north. The butterflies are responding effectively to climate change—which is certainly what we would hope a butterfly species would do on a planet with a climate history as variable as Earth's.

Parmesan's study found populations of Edith's checkerspot butterflies thriving over most of western North America. She found fewer of them at the southernmost extremity of their range—in Baja California and near San Diego—than in the past. However, huge expanses in Canada have been warming into their preferred climate range.

Finally, we must recognize that species resist extinction strongly and they often persist even when humans think they've been wiped out. The supposedly extinct ivory-billed woodpecker has recently been confirmed as living in two forests in eastern Arkansas.[55] The Nature Conservancy has recently found three "extinct" snails in Alabama and California botanists have found a plant called the Mount Diablo buckwheat for the first time since 1936. At least twenty-four other "extinct" species have been found during natural heritage surveys in North America since 1974.[56]

Raising the Alarm about Coral Reefs:

"Corals will inevitably be among the first organisms to show the consequences of a sustained increase in sea surface temperatures because of the fragile temperature dependence of the tiny algae, called zooxanthellae, which live in the coral's cells. The coral's color and most of its food come from these algae, so without them the coral cannot grow."[57]

"A rise in water temperature in the Indian Ocean by just one degree over a few weeks in 1998 destroyed 80 to 90 percent of the coral," [Mark] Spalding says. "Whole reefs around the Maldives and Seychelles pretty much died."[58]

"The Philippine coral reefs, among the most diverse and largest in the world, may not be around for long. . . . On the last day of the symposium [at Bali] the environmental group Greenpeace released a new coral reef study showing that, because of global warming, the Pacific Ocean could lose most of its coral reefs by the end of the current century."[59]

ARE BLEACHED CORAL DYING OR HOUSECLEANING?

The claim that higher temperatures will kill off the world's corals is irresistible to global warming activists. They understand the emotional appeal of the reefs and their bright-colored fishes. The only problem is that the higher-temperatures-kill-corals theory is literally incredible.

Corals date back 450 million years, and most of today's coral species date back at least 200 million years. Just in the last two million years, coral reefs have been through at least seventeen glacial periods, interspersed with their warm interglacial periods. These glacial-interglacial shifts imposed repeated dramatic temperature changes, along with sea level changes as drastic as four hundred feet (during the last ice age).

Temperature changes across the Pacific are related to the El Niño–South Oscillation (ENSO), which causes a major Pacific temperature change every four to seven years. The 1998 El Niño boosted sea surface temperatures all over the Pacific, causing massive coral bleaching, especially in the Indian Ocean. That's when Mark Spalding, supposedly a coral expert, claimed that the vast majority of the corals had died out.

But wait. Bleaching is how corals adapt to changing temperatures.[60]

R. J. Jones of Australia's Queensland University reported coral bleaching on a portion of the Great Barrier Reef just after average daily sea temperatures rose by 2.5 degrees in eight days.[61] However, Canada's D. R. Kobluk and M. A. Lysenko found severe coral bleaching in the Caribbean after the water *declined* 3 degrees Celsius in eighteen hours.[62]

New studies tell us that bleaching is the coral's system for dealing with sudden temperature changes.

Cynthia Lewis and Mary Alice Coffroth of the University of Buffalo deliberately triggered bleaching in some coral colonies. In response, the colonies ejected 99 percent of their symbiotic algae friends. The researchers then exposed the bleached coral to a rare variety of algae that wasn't in the coral colonies at the beginning of the experiment. Sure enough, within a few weeks, the corals had substantially restocked their algae shelves, and about half included the new marker algae. Later, the marker variety was displaced from several of the coral colonies by more effective algae strains—indicating that the corals pick the best partners for the new conditions from the whole variety of algae floating in their part of the ocean.[63]

Lewis and Coffroth say this is a healthy demonstration of flexibility in coral colonies. They say the coral systems have the flexibility to establish new associations with algae strains from the whole environmental pool and that this is "a mechanism for resilience in the face of environmental change."[64]

In the same 2 June 2004 issue of *Science*, Angela Little of Australia's James Cook University and Madeline van Oppen of Australia's Institute of Marine Science echoed the findings of Lewis and Coffroth. They attached tiles to various parts of the Great Barrier Reef in Australia and studied the algae recruited by the juvenile corals that grew on them. They found that the young corals were likely to try any algae—"a potentially adaptive trait."[65]

Eco-activists and biologists who claim global warming is killing the coral are wrong—again. However, science has already returned its verdict on coral and global warming: No link.

CONCLUSION

We have looked at the theory and the real-world evidence of mass extinction due to global warming. We found no persuasive rationale that huge numbers of species would die from global warming, nor any real-world wild species losses due to the warming the Earth has recently experienced. Instead, we have found a great deal of evidence that species move effectively to keep or expand their ranges in response to the climate change characteristics of our planet. By their movements, they are effectively testifying against the rigidity of the Greenhouse Theory and for the flexibility inherent in the 1,500-year cycle.

We have also found both theory and evidence that higher concentrations of CO_2 help plants—and thus ultimately other species—to accept higher temperatures without harm, or even energy loss. That the extinction theorists continue to ignore this is inexcusable. Wildlife biologists probably hate to take advice from colleges of agriculture, where most of the CO_2 research has been done. However, the CO_2 research is nevertheless a vital element of global warming threat analysis.

And, we have found a claim of a species loss to global warming (the Golden Toad) that should not have been offered without a caveat, either by the authors or by *Nature*'s editors; both should have known about the deforestation study that seems to refute the claim that sea surface warming caused the disappearance of the Golden Toad. We have found a reputable biologist, Chris Thomas, making scary claims about massive extinctions that are refuted by his own published research. We have found a well-known biologist, Camille Parmesan, authoring a poorly supported and highly overstated claim, in a prestigious scientific journal, and repeatedly misusing the term "locally extinct" to overdramatize climate warming risks. Finally, we find eco-activists and biologists who claim global warming is killing the coral, while science has already returned its verdict on coral and global warming: No link.

The public is not being well served. We are being humbugged by activists with no credentials, of whom we should automatically be wary. We are also being humbugged by the journalists we pay to provide us with reliable information, healthy skepticism, and differing perspectives. Now we are even being humbugged by highly trained professional scientists, many of them working on government research grants.

NOTES

1. Guy Gugliotta, "Mass Extinction Looms by 2050, Climate Study Finds," Washington Post, 8 January 2004.

2. Julie Deardorff, "Studies Show Global Warming Is Affecting Plant, Animal Species," Chicago Tribune, 3 January 2003.

3. Guy Gugliotta, "Impact Crater Labeled Clue to Mass Extinction," Washington Post, 14 May 2004.

4. C. D. Thomas et al., "Extinction Risk from Climate Change," *Nature* 427 (2004): 145–48.

5. T. Root et al., "Fingerprints of Global Warming on Wild Animals and Plants," *Nature* 421 (2003): 57–60; C. Parmesan and G. Yohe, "A Globally Coherent Fingerprint of Climate Change Impacts across Natural Systems," *Nature* 421 (2003): 37–42.

6. J. Levinton, "The Big Bang of Animal Evolution," *Scientific American* 267 (1992): 84–91.

7. P. F. Schuster et al., "Chronological Refinement of an Ice Core Record at Upper Fremont Glacier in South Central North America," *Journal of Geophysical Research* 105 (2000): 4657–666.

8. Guy Gugliotta, "Impact Crater Labeled Clue to Mass Extinction," *Washington Post*, 14 May 2004.

9. Jared Diamond, *Guns, Germs and Steel* (New York: W. W. Norton, 1997), 46–47.

10. Dennis T. Avery, *Saving the Planet with Pesticides and Plastics: The Environmental Triumph of High-Yield Farming* (Indianapolis, IN: Hudson Institute, 2000), 36–37.

11. C. D. Thomas et al., "Extinction Risk from Climate Change," 145–48.

12. J. A. Pounds et al., "Biological Response to Climate Change on a Tropical Mountain," *Nature* 398 (1999): 611–15.

13. J. A. Pounds and S. H. Schneider, "Present and Future Consequences of Global Warming for Highland Tropical Forests Ecosystems: The Case of Costa Rica," paper presented at the U.S. Global Change Research Program Seminar, Washington, D.C., 29 September 1999.

14. R. O. Lawton et al., "Climatic Impact of Tropical Lowland Deforestation on Nearby Mountain Cloud Forests," *Science* 294 (2001): 584–87.

15. Lawton et al., "Climatic Impact of Tropical Lowland Deforestation on Nearby Mountain Cloud Forests," 584–87.

16. J. A. Neeley and S. J. Scherr, *Common Ground, Common Future* (Washington, D.C.: The World Conservation Union/Future Harvets, 2001), 1–23.

17. Norman Borlaug, press release, Declaration in Support of High-Yield Conservation, Washington, D.C., 30 April 2002.

18. C. Loehle, "Height Growth Rate Tradeoffs Determine Northern and Southern Range Limits for Trees," *Journal of Biogeography* 25 (1998): 735–42.

19. Ibid. See also Y. Gauslaa, "Heat Resistance and Energy Budget in Different Scandinavian Plants," *Holarctic Ecology* 7 (1984): 1–78; J. Levitt, *Responses of Plants to Environmental Stresses, Vol. 1: Chilling, Freezing and High Temperature Stresses* (New York: Academic Press, 1980); L. Kappen, "Ecological Significance of Resistance to High Temperature," in *Physiological Plant Ecology. I. Response to the Physical Environment*, ed. O. L. Lange et al. (New York: Springer-Verlag, 1981), 439–74.

20. C. Loehle, "Height Growth Rate Tradeoffs," 735–42

21. S. Idso, C. Idso, and K. Idso, *The Specter of Species Extinction* (Washington, D.C.: The Marshall Institute, 2003), 1–39.

22. Ibid. See also D. I. Axelrod, "The Oakdale Flora (California)," *Carnegie Institute of Washington Publication* 553 (1944): 147–200; D. I. Axelrod, "Mio-Pliocene Floras from West-Central Nevada," *University of California Publications in the Geological Sciences* 33 (1956): 1–316; D. I. Axelrod, "The Late Oligocene Creede Flora, Colorado," *University of California Publications in the Geological Sciences* 130 (1987): 1–235.

23. Idso, Idso, and Idso, *The Specter of Species Extinction*, 1–39.

24. K. E. Idso and S. B. Idso, "Plant Responses to Atmospheric CO_2 Enrichment in the Face of Environmental Constraints: A Review of the Past 10 Years' Research," *Agriculture and Forest Meteorology* 69 (1994): 153–203.

25. M. G. R. Cannell and H. H. M. Thornley, "Temperature and CO_2 Responses of Leaf and Canopy Photosynthesis: A Clarification Using the Nonrectangular Hyperbola Model of Photosynthesis," *Annals of Botany* 82 (1998): 883–92.

26. R. R. Nemani et al., "Climate-Driven Increases in Global Terrestrial Net Primary Production from 1982 to 1999," *Science* 300 (2003): 1560–563.

27. B. A. Kimball, "Carbon Dioxide and Agricultural Yield: An Assemblage and Analysis of 430 Prior Observations," *Agronomy Journals* 75 (1983): 779–88; K. E. Idso and S. B. Idso, "Plant Responses to Atmospheric CO_2 Enrichment in the Face of Environmental Constraints," 153–203; and H. Saxe et al., "Tree and Forest Functioning in an Enriched CO_2 Atmosphere," *New Phytologist* 139 (1998): 395–436.

28. T. Root et al., "Fingerprints of Global Warming on Wild Animals and Plants," *Nature* 421 (2003): 57–60.

29. C. Parmesan and G. Yohe, "A Globally Coherent Fingerprint of Climate Change Impacts across Natural Systems," *Nature* 421 (2003): 37–42.

30. P. Hersteinsson and D. W. Macdonald, "Interspecific Competition and the Geographical Distribution of Red and Arctic Foxes, *Vulpes vulpes* and *Alopex lagopus*," *Oikos* 64 (1992): 505–15.

31. K. Taira, "Temperature Variation of the 'Kuroshio' and Crustal Movements in Eastern and Southeastern Asia 700 Years B.P.," *Palaeogeography, Palaeoclimatology,*

Palaeoecology 17 (1975): 333–338; A. M. Korotky et al., *Development of Natural Environment of the Southern Soviet Far East (Late Pleistocene-Holocene)* (Moscow, Russia: Nauka, 1988).

32. C. D. Thomas et al., "Ecological and Evolutionary Processes at Expanding Range Margins," *Nature* 411 (2001): 577–81.

33. Thomas et al., "Ecological and Evolutionary Processes at Expanding Range Margins," 577–81.

34. Thomas et al., "Ecological and Evolutionary Processes at Expanding Range Margins," 577–81.

35. Thomas et al., "Ecological and Evolutionary Processes at Expanding Range Margins," 577–81.

36. Jeffrey Levinton, "The Big Bang of Evolution," *Scientific American* 267 (November 1992): 84.

37. Simon Collins, "Antarctic Fish Set to Survive Warmer Seas," *New Zealand Herald*, 16 April 2004.

38. Elizabeth Day, "Charities 'Spread Scare Stories on Climate Change to Boost Public Donations,'" *Weekly Telegraph/UK*, 24 February 2005.

39. R. J. Ladle, P. Jepson, M. B. Araújo, and R. J. Whittaker, "Dangers of Crying Wolf over Risk of Extinctions," *Nature* 428 (22 April 2004): doi: 10.1038/428799b.

40. Parmesan and Yohe, "Globally Coherent," 37–42.

41. C. D. Thomas and J. J. Lennon, "Birds Extend Their Ranges Northwards," *Nature* 399 (1999): 213.

42. G. Grabherr et al., "Climate Effects on Mountain Plants," *Nature* 369 (1994): 448.

43. H. Pauli et al., "Effects of Climate Change on Mountain Ecosystems—Upward Shifting of Mountain Plants," *World Resource Review* 8 (1996): 382–90.

44. E. S. Vesperinas et al., "The Expansion of Thermophilic Plants in the Iberian Peninsula as a Sign of Climate Change," in *"Fingerprints" of Climate Change: Adapted Behavior and Shifting Species Ranges*, ed. G. Walther et al. (New York: Kluwer Academic/Plenum Publishers, 2001), 163–84.

45. C. M. Van Herk et al., "Long-Term Monitoring in the Netherlands Suggests that Lichens Respond to Global Warming," *Lichenologist* 34 (2002): 141–54.

46. E. Pollard et al., "Population Trends of Common British Butterflies at Monitored Sites," *Journal of Applied Ecology* 32 (1995): 9–16.

47. A. J. Southward, "Seventy Years' Observations of Changes in Distribution and Abundance of Zooplankton and Intertidal Organisms in the Western English Channel in Relation to Rising Sea Temperatures," *Journal of Thermal Biology* 20, no. 1 (February 1995): 127–55.

48. R. C. Smith et al., "Marine Ecosystem Sensitivity to Climate Change," *BioScience* 49 (1999): 393–404.

49. R. I. L. Smith, "Vascular Plants as Bioindicators of Regional Warming in Antarctica," *Oecologia* 99 (1994): 322–28.

50. R. D. Sagarin et al., "Climate-Related Change in an Intertidal Community over Short and Long Time Scales," *Ecological Monographs* 69 (1999): 465–90.

51. M. Sturm et al., "Increasing Shrub Abundance in the Arctic," *Nature* 411 (2001): 546–47.

52. N. K. Johnson, "Pioneering and Natural Expansion of Breeding Distributions in Western North American Birds," *Studies in Avian Biology* 15 (1994): 27–44.

53. Kevin Krajick, "All Downhill from Here," *Science* 303 (2004): 1600–602.

54. Krajick, "All Downhill from Here," 1600–602.

55. Arkansas Game and Fish Commission, "Ivory-Billed Woodpecker Found in Arkansas," 5 August 2005.

56. Mark Schaefer of NatureServe, an Arlington, Virginia, conservation group, quoted in "When Extinct Isn't," 8 August 2005, <ScientificAmerican.com>.

57. Greenpeace, "Bleaching Corals," 1994, <http://archive.greenpeace.org/climate/ctb/corals.html> (accessed 26 February 2004).

58. Samantha Sen, "Disappearing Coral Reefs Also a Human Loss," Interpress News Service, 13 September 2001.

59. Report from the 9th International Coral Reef Conference, Bali, 23 June 2003, <www.codewan.com.ph/CyberDyarayo/features/f2000_1113_01.htm>.

60. Richard Black, "Coral Reefs Adapting to Rising Coral Temperatures?" Cyber Diver News Network, <www.cdnn.info/news/eco/e060607a.html> (7 June 2006).

61. R. L. Jones et al., "Changes in Zooxanthellar Densities and Chlorophyll Concentration in Corals during and after a Bleaching Event," *Marine Ecology Progress Series* 158 (1997): 51–59.

62. D. R. Kobluk and M. A. Lysenko, "Ring Bleaching in Southern Caribbean Agaricia Agaricites during Rapid Water Cooling," *Bulletin of Marine Science* 54 (1994): 142–50.

63. D. L. Lewis and M. A. Coffroth, "The Acquisition of Exogenous Algal Symbionts by an Octocoral after Bleaching," *Science* 304 (2004): 1490–491.

64. Lewis and Coffroth, "The Acquisition of Exogenous Algal Symbionts by an Octocoral After Bleaching," 1490–491.

65. A. F. Little et al., "Flexibility in Algal Endosymbioses Shapes Growth in Reef Corals, Science 304 (2004): 1492–494; Kobluk and Lysenko, "Ring Bleaching in Southern Caribbean Agaricia Agaricites during Rapid Water Cooling," 142–50.

Chapter Seven

Warming and Cooling
in Human History

Harkening to the Human Histories:

"The snow began to fall about the 10th of November 1779, and continued falling almost every day until the middle of the following March. The Northeast virtually shut down. It was known as The Hard Winter. . . . It was a world of Ice. . . . Judge Jones, who lived at Fort Neck (now Massapequa), wrote in his book that 200 provision-laden sleighs, pulled by two horses each, escorted by 200 light cavalry, made the five-mile trip from New York to Staten Island. . . . The British hauled cannon across the ice from Manhattan to defend themselves. . . . The *New York Packet* reported a thermometer reading of 16 below zero in the city. . . . The severe cold reached up and down the coast from Maine to Georgia."[1]

"Viking travels in North America can be traced thanks to a geological feature in [modern Labrador] that Leif the Lucky called the Furdustrand ("Wonder Strand"), a beach of white sand bordered by dense woodland forty miles long and in places 200 feet wide. The beach still exists, and references to it in the two pertinent Viking sagas reveal a fairly accurate itinerary. By this means, the major Viking settlement was located at L'Anse-aux-Meadows in 1960. It was there that the Viking found grapes. . . . Nothing brings home the effects of Warming more than the fact that Newfoundland was once called "Vinland."[2]

"Glacier advances in the vicinity of Mont Blanc, France, destroyed three villages and heavily damaged a fourth between 1600 and 1610. The oldest of these villages had existed since the 1200s."[3]

CLIMATE IS HARD TO SEE

Between the full-scale, 100,000-year ice ages, the Earth's climate has been dominated by a moderate, natural, irregular 1,500-year cycle. The most recent examples of the cycle include the Roman Warming, which started about 200 B.C. and paired with its other half, the Dark Ages, ended about A.D. 900. The Medieval Warming, paired with its other half, the Little Ice Age, together lasted from 900 to 1850. Is our own Modern Warming (1850–present) the first half of the next cycle?

The people who lived during these cycles were not aware of them. They thought of themselves as enjoying good weather or suffering bad weather. The 1,500-year length of the cycle was too long for humans to understand before written records and too moderate to perceive without accurate thermometers. Only recently, with gleanings from written histories going back to before the Roman Warming and with the physical evidence in ice cores, tree rings, and other proxies, have we begun to understand the long, moderate cycle that governs our interglacial climate.

We have lately been conditioned by global warming headlines to think of climate change as a dramatic event. In reality, watching global warming—or even global cooling, a far more drastic development—is mostly less exciting than watching grass grow. The change comes primarily in the form of marginally higher or lower temperatures in the middle of the winter, at night. It can come in the form of more or fewer storms—but storms are always a part of life on Earth. At the equator, climate change is reflected in rainfall changes, but there are always droughts and floods.

Only a few people actually saw their Alpine villages crushed by the advancing ice of a giant glacier during the Little Ice Age. Only a few people caught quick glimpses of the Inuit kayaks off the coast of Scotland.

The Slavs who moved to the upper hillsides of the Eastern Alps in the latter part of the Medieval Warming didn't realize that the land had been abandoned during the last cooling. They probably thought the earlier German settlers had just gotten too soft and affluent to keep climbing that far uphill. They certainly had no idea that the climate around them was about to make their newly settled farms suddenly untenable again.

Even "abrupt" climate change can take a century to get the public's attention. Our Modern Warming began about 1850 but few people noticed until after the surge of warming that occurred between 1920 and 1940.

Through it all, the world's communities continued to experience harsh cold, searing heat, heavy rains and/or snows and crop-killing droughts. There continued to be hurricanes, cyclones, blizzards, ice storms, and floods.

The Romans lived through one of the important warmings with almost no notice of it in their voluminous writings. Only the gradual northward progression of grape vines (in both Italy and Britain) and olive trees (in Italy) seems to have registered the Roman Warming in their minds.

By the time the Romans had to contend with the global cooling of the Dark Ages, they also had to contend with the invading barbaric tribes. Neither the Romans nor barbarians would have been thinking much about long-term climate cycles as their swords and spears clashed against each others' shields outside the empire's city walls. It is only today that we can understand that the spreading drought and accompanying famine of the cooling climate was a factor in driving the Central Asian barbarians to attack the awesome Roman gates.

The Icelanders lived through the global warming/cooling cycle—between 900 and the end of the Little Ice Age in 1850—in one of the most climate-vulnerable communities on the planet. Books and official records survive from this period, and sagas supplied even earlier history commentary. Yet, until about the year 1920, the Icelanders apparently satisfied themselves that they had been dealing only with periodic bad weather, not long-term climate change.

Even when our forebears didn't recognize the climate was changing, however, the climate changes were recorded in such places as monastery chronicles, the accounts of feudal estates, legal papers, government reports, harbor records, and the myriad writings of travelers, scholars, and nobles. These will be key resources for our trip back in time.

CIRCA 750 B.C. TO 200 B.C.:
THE COLD BEFORE THE ROMAN WARMING

Egyptian records document an unnamed cooling trend about 750 to 450 B.C.—just before the founding of Rome. The Egyptians had to build dams and canals to deal with a decline in the beneficial Nile floods resulting from the cooler, drier climate. The decreasing flood levels are also recorded in Central Africa, where sediments show that the level of Lake Victoria progressively declined.

Early Roman authors wrote of a frozen Tiber River and of snow remaining on the ground for lengthy periods. Those events would be unthinkable today. We also now know that European glaciers advanced during the early part of the Roman civilization.

"The resulting drop in sea level [due to more water being trapped in glaciers and ice sheets] is borne out by many cultural features. Port facilities . . . were constructed and conditioned to a sea level lower than that of today.

Evidence from the Mediterranean area is also found in Egypt, where the Sweetwater Canal, built as a response to lowered sea level, became silted," says John E. Oliver, author of *Climate and Man's Environment*.[4]

200 B.C. TO A.D. 600: THE ROMAN WARMING

After the first century B.C., Romans were writing of a warmer climate, with little snow or ice, and with grapes and olives growing farther north in Italy than had been possible in earlier centuries.

"By 350 A.D.," Oliver concludes, "the climate had become milder in northern realms while in tropical regions it appears to have become excessively wet. Tropical rains in Africa caused high-level Nile floods and temples built earlier . . . were inundated. At this time, too, Central America experienced heavy precipitation and tropical Yucatan was very wet."[5]

Hubert H. Lamb, who founded the Climatic Research Center at Great Britain's University of East Anglia, was the first to document the changes in the world's climate over the past 1,000 years. He says the Romans reported grapes first being cultivated at Rome about 150 B.C. Other reports also seem to confirm that through the Roman times the Italian and European climate was becoming warmer; there was "a recovery from a period around 500 B.C. which was colder than now. By late Roman times, particularly in the fourth century A.D., it may have been warmer than now."[6]

Columella, a Roman who wrote during A.D. 30–60 quoted earlier writers who said that in their time "the vine and the olive were still slowly working their way northwards up the leg of Italy. . . ." Columella added that "Saserna . . . concludes that the position of the heavens has changed . . . *regions which previously, on account of the regular severity of the weather, could give no protection to any vine or olive stock planted there, now that the former cold has abated and the weather is warmer, produce olive crops and vintages in great abundance*"[7] (emphasis added).

"The Mediterranean climate itself differed from today," says Oliver. "Weather records kept by Ptolemy in the second century A.D. show that precipitation occurred throughout the year, in contrast to the winter maximum of today."[8]

Rome also experienced a moderate rise in sea level as the warming began. Lamb estimates it at one meter or less, as more of the world's glacial ice melted. Remains of ancient harbors at Naples and in the Adriatic are now three feet below current water levels.[9]

North Africa (what's now Tunisia, Algeria, and Morocco) was moist enough to grow large amounts of grain, first for the Carthaginians and then for the Roman Empire.

Central Asia experienced strong population growth when the climate there began to warm around A.D. 300, according to Robert Claiborne.[10]

A.D. 440 TO A.D. 900: THE DARK AGES

A major climate catastrophe ushered in the Dark Ages. Professor Michael Baillie of Queens University in Northern Ireland says tree ring studies verify the major cooling. Moreover, he points to a particular period that he calls the "A.D. 540 event" when "the trees stopped growing." He discovered that flooded bog oaks and timber from this era display narrow rings indicating that trees suddenly and inexplicably stopped growing around the world. Temperatures dropped. He notes that snow fell during the summer in southern Europe and in coastal China while savage storms swept from Sweden to Chile.[11]

"The trees are unequivocal that something quite terrible happened," asserts Baillie. "Not only in Northern Ireland and Britain, but right across northern Siberia, North and South America, a global event of some kind [occurred].

"There seem to have been comets, meteors, earthquakes, dimmed skies and inundations and, following the famines of the late 530s, plague arrived in Europe in the window AD 542–545," notes Baillie.[12]

Byzantine historians recorded many frightening comets in the sky. This is not just mentioned in passing, since "close call" comets can raise enough cosmic dust to temporarily change the earth's climate patterns. The historian Procopius, describing the weather in Constantinople at that time, reported that "[t]he sun gave forth its light without brightness, like the moon during the whole year and it seemed exceedingly like the sun in eclipse, for the beams it shed were not clear, nor such as it is accustomed to shed."[13] John of Ephesus wrote, "The sun became dark and its darkness lasted for 19 months. Each day it shone for about four hours, and still this light was only a feeble shadow . . . the fruits did not ripen and the wine tasted like sour grapes."[14] John the Lydian wrote in *De Ostentis*: "The sun became dim for nearly the whole year . . . the fruits were killed at an unseasonable time."[15]

"At the close of the eighth century," says Oliver, "evidence points to a marked cooling. In 800 to 801 A.D., the Black Sea was frozen and in 829 A.D., ice formed on the Nile."[16]

Disease and Warfare Follow Drought and Hunger

Could Asian droughts between 300 and 800 have driven the barbarian tribes from central Europe and Asia to attack the West? Germans call this period the *Volkerwanderungen*, the "folk wandering time," because Eastern Europe was invaded by so many of the "folk wandering."

The first huge epidemic of bubonic plague probably came out of the Middle East or East Africa (it resides in both places). Perhaps triggered by drought, it was spread by rats and their fleas, transmitting the deadly *Yersinia pestis* bacterium to new groups of humans who had no resistance. The plague killed perhaps 200,000 citizens in the Byzantine Empire in four months, then went on to kill about one-third of the people in Eastern Europe and half the population of Western Europe. The total plague death toll may have been 25 million.

Trade came to a halt. Rats traveled on ships, and sailors were suspected of spreading the disease. In some port cities they were not allowed off their ships—but the rats and fleas leaped ashore to spread the plague anyway. "It appears that the plague happened as a result of the famine caused by the lack of sunlight," writes Byzantine historian Cyril Mango.[17]

It was called "Justinian's Plague," because it occurred just as the Emperor Justinian was attempting to rebuild the old Roman Empire from his base in Constantinople. He was even trying to reconquer Italy from the barbarian tribes. Justinian's Plague was seen as God's condemnation of the Romans—both by the Romans themselves and by the neighboring Persians and Arabs. Weakened by the plague, Rome's former eastern provinces fell victim to rapid Islamic expansion after Mohammed's death in 632.

After Justinian's Plague had run its course about 590, the disease disappeared from view until the 1320s—another climatic turning point when the Medieval Warming was shifting into the Little Ice Age. Temperatures fell, huge storms swept across the Atlantic, and drought assailed Asia.

900 TO 1300: THE MEDIEVAL WARMING

We have known for a long time that the world had very favorable weather during the medieval period, from perhaps 900 to 1300. The period is known to history as the Medieval Warming or the Little Climate Optimum. (The much-warmer and longer Holocene Climate Optimum occurred between 5,000 and 9,000 years ago.)

Many of the famous castles and cathedrals of Europe were built during the Medieval Warming, indicating good crops, ample food supplies, and enough off-farm labor to undertake major construction projects.

Hubert H. Lamb says that "[w]here there is no reasonable doubt is that, over the next three to four centuries, [after A.D. 800] . . . we see that the climate was warming up, until there came a time when cultivation limits were higher up on the hills than they have ever been since. . . . Certainly the upper tree line in parts of Central Europe was 100 to 200 meters higher than it became by the seventeenth century. . . . On the heights in California, the tree ring record indicates that there was a sharp maximum of warmth, much as in Europe, between A.D. 1100 and 1300."[18]

Norse sagas and written records tell us it was during this warm period that the Norse colonized Greenland. The colonists supported themselves by catching codfish and hunting seals in ice-free seas, and pasturing cattle and sheep at sites that then became frozen tundra for five hundred years.

Richard D. Tkachuck, of the Geosciences Research Institute, writes:

Human remains in Norse burial grounds located in Greenland have been found which are now in permanently frozen soils. This suggests an average local temperature at the time of Norse occupation 2–4 degrees C higher than at present. Additionally, the finding of plant roots at this same level supports this supposition, since the permafrost layer provides a barrier to growth. There is evidence that American Eskimos occupied areas in the north of Greenland, on Ellesmere Island and the New Siberia Islands. At these locations, large dwellings made from driftwood have been found. There is also archeological evidence of large villages that were developed for whaling and fishing. These settlements eventually were forced south by climatic change until they came in contact with Viking colonies in southern Greenland. Conflict occurred, and the Viking colonies eventually died out in the 1400s.[19]

The Vikings named Newfoundland "Vinland." As noted earlier, grape vines are one of the key temperature proxies for the Warming and Cooling of the last 1,200 years. England's Domesday Book, for example, indicates widespread growth of grape vines in areas where grapes are now not suitable. Thus, northwestern Europe must have been warmer and drier than it is now.[20]

Tkachuck confirms:

The cultivation of grapes for wine making was extensive throughout the southern portion of England from about 1100 to around 1300. This represents a northward latitude extension of about 500 km from where grapes are presently grown in France and Germany. . . . With the coming of the 1400s, temperatures became too cold for sustained grape production, and the vineyards in these northern latitudes ceased to exist. . . . *It is interesting to note that at the present time the climate is still unfavorable for wine production in these areas.* . . . In this warm time, vineyards were found at 780 meters above sea level in Germany. Today they are found up to 560 meters. If one assumes a 0.6 to 0.7° C. change/100 meter vertical excursion, these data imply that the average mean temperature was 1.0 to 1.4° C. higher than the present[21] (emphasis added).

Brian Fagan describes the impact of the Medieval Warming in his book about the Little Ice Age:

> Settlement, forest clearance and farming spread 100 to 200 meters farther up valleys and hillsides in central Norway, from levels that had been static for more than 1,000 years. Wheat was grown around Trondheim [almost halfway up the Scandinavian Peninsula], and hardier grains such as oats as far north as Malagan [even closer to the Arctic Circle]. . . . The height change hints at a rise in summer temperatures of about a degree Centigrade, a similar increase to that across the North Sea in Scotland. . . . During the late prehistoric times, numerous copper mines had flourished in the Alps until the advancing ice sealed them off. Late medieval miners reopened some of the workings when the ice retreated.[22]

Archaeologists digging under the northern English city of York have found remains of the nettle groundbug, *Heterogaster urticae*, whose typical habitat today is on stinging nettles at sunny locations in the south of England. It was present much farther north, in York, both during the Roman era and the Middle Ages. Obviously, temperatures must have been higher than today in both those periods.[23]

Climate and Food

The biggest climate impact on most people was through food production. Europe's population increased about 50 percent during the Medieval Warming. That meant food production—or at least the certainty of relatively good crops year in and year out—must have increased even more than that. Population numbers would have been governed by the food supply in the poorer crop years. Food stocks held for the next year were never huge, given the primitive storage bins and the pervasive abundance of rats and chewing insects in that prepesticide era.

The food abundance of the Medieval Warming ensured the growth of population, cities, transportation, and great buildings. Even a one-degree difference in the winter climate made a significant difference in the length of the growing season, especially the absence of untimely frosts.

During June 1253, Westminster Abbey alone had 428 construction workers, nearly half of them skilled stone workers and glass blowers, and there were hundreds of other major building projects across Europe.[24]

Since there were only one-tenth as many people in Europe then as today, it would be comparable to a modern public monument project employing 4,280 construction workers—for decades.

Population Cycled with Climate

In the late seventeenth century, Gregory King estimated the British population at 5.5 million thanks to new farming systems that produced more food despite the relatively poor weather of the Little Ice Age. British agricultural historian Mark Overton says it had probably reached that level twice before: (1) during the warming of the Roman period and (2) in 1300, at the height of the Medieval Warming.[25]

Working from records of the land held in trust by probate courts, Overton shows the productivity of British grain lands dropped significantly between 1300 and 1400 as the climate deteriorated. The yields did not recover until about 1700 when Britain's farmers adopted crop rotation, with its higher yields.[26]

Warming Favored Travel and Trade

Overseas and coastal trade was encouraged during the Medieval Warming by a decline in the number of heavy winds and fierce storms encountered by ships. Roads benefited from lots of sunshine; even if it rained often, the roads quickly dried out so that carts could carry food to the cities and city goods back to the farms. Mountain passes stayed open longer in the summers, so that luxury goods such as spices from the Oriental caravans, sugar from Cyprus, and fancy glass from Venice could be traded for English woolens and Scandinavian furs.

Medieval weather was warm and dry enough to encourage the emergence of Europe's first trade fairs. In the year 1000, Britain had more than seventy mints scattered among its market towns, coining German silver received in payment for British wool and fish. There was also European trade in wine, furs, cloth, and slaves.[27] Earlier, the economy had been organized around self-sufficient estates, which grew their own food, flax, and wool; wove their own clothing; and traded little with anyone.

Medieval Warming around the Globe

The Mediterranean Region

This region, including the coast of North Africa, got more rainfall in those days than it does in our time. Farther south, in the desert regions of North Africa, the writings of the great Arab geographers indicate there was more rainfall than now all through the Middle Ages.[28]

Asia

The Chinese climate from 1000 B.C. to A.D. 1400 has been reconstructed from palace records, official histories, yearbooks, diaries, and gazettes.[29] Key elements include the arrival dates of migrating birds; distribution of plant species, bamboo groves and fruit orchards; patterns of elephant migrations; the flowering dates of shrubs; and the major floods and droughts.

The records affirm other evidence of longer growing seasons and a warmer climate. Chinese citrus groves moved north with the Medieval Warming, and then retreated south when beset by the cooling after 1300.[30]

Chinese temperatures were 2 to 3 degrees Celsius higher than present during China's Climate Optimum (8000 B.C. to 3000 B.C.) based on pollen data, and the southern limits of permafrost.[31]

Chinese wealth, which had been rising since 200 B.C., peaked about A.D. 1100, and then declined, according to Kang Chao's careful economic analysis of that country's history. Chao says real earnings rose from the Han period (206 B.C. to A.D. 220) to a peak during the Northern Sung Dynasty (961 to 1127).[32]

Y. Tagami assembled sets of historic Japanese documents, including a set of official records dating from the seventh century on weather "events" such as droughts, long rains, heavy snows, mild winters, and the like.[33] Other types of documents, including diaries of Japanese nobles, recorded weather events related to their social calendars, including cherry blossom viewing dates and lake freezing dates. This set of documents extends back to the tenth century.

Tagami also analyzed the number of days with snowfall relative to days with rainfall:

> Relatively hot conditions continued until the eighth century, [then] cool conditions appeared for a short period in the late ninth century. Then warm conditions continued from the 10th century to the former half of the 15th century. After the latter half of the 15th century, cool conditions appeared and then considerable cold conditions started from the 17th century. So, between the former and the latter cold ages, the warm condition is clear from the 10th century to the 14th century.[34]

In Southeast Asia, the thousands of famous temples at Angkor Wat were built during the first half of the twelfth century, probably reflecting the same favorable weather for food production and labor force expansion that was registered by the castles and cathedrals built in Europe.

North America

In the southwest, the peaceful Anasazi Indians lived on individual scattered water-gathering farms across the western Colorado plateau and the surrounding parts of northern Arizona and New Mexico. The Anasazi community's corn and squash prospered, and their culture spread, from perhaps 2000 B.C. to at least A.D. 1000.

The early part of the Medieval Warming favored the Anasazi with more and more consistent rainfall. However, tree rings in Sand Canyon show little rain from 1125 to 1180, from 1270 to 1274 and a twenty-four-year drought in the latter thirteenth century. Prolonged drought made food scarce, and triggered conflict.

The Anasazi were forced to build the fortress-like cliff-side dwellings at Mesa Verde and Chaco Canyon. Eventually, even the cliff dwellings were attacked and the grain stores burned or looted. After 1300, the remnants of the Anasazi found their way to the few river valleys (such as the Rio Grande and Colorado) where water for irrigation still ran.[35]

Farther east, the corn-based Indian culture of the mound-builders had flourished in what is now the Eastern Corn Belt. Cahokia, Illinois, is estimated to have had 40,000 inhabitants before 1200. But the cooler, drier climate after 1200 made it harder to grow corn. The local trees were gradually displaced by grasses that needed less rainfall, and the deer and elk gave way to bison, which were hard to kill without the horses brought in later by the Spaniards. Without corn, Cahokia could not be supported. When the first French traders arrived in the eighteenth century, they found only scattered Indian villages.[36]

1300 TO 1550: THE LITTLE ICE AGE—PHASE ONE

One of the most important benefits of the Medieval Warming was that the climate was relatively stable. People and society could successfully adapt to it. One of the greatest problems with the Little Ice Age was the instability of the climate. Archaeologist Brian Fagan says the variability of winter temperatures in England and the Netherlands was about 40 to 50 percent greater during the coldest centuries of the Little Ice Age than during the early twentieth century.[37]

It was virtually impossible to adapt and all too necessary to suffer.

During the Little Ice Age, the climate was unpredictable with often-wild weather delivering warm and very dry summers in some years, very cold and wet ones in other years, and with a notable increase in storms and winds in the North Sea and English Channel.

According to Fagan:

A modern European transported to the height of the Little Ice Age would not find the climate very different, even if winters were sometimes colder than today and summers very warm on occasion. There was never a monolithic deep freeze, rather a climatic seesaw that swung constantly backwards and forwards. . . . There were Arctic winters, blazing summers, serious droughts, torrential rain years, often-bountiful harvests, and long periods of mild winters and warm summers. Cycles of excessive cold and unusual rainfall could last a decade, a few years or just a single season.[38]

The first advancing glaciers, the unmistakable signal that the Little Ice Age was arriving, began in Greenland during the early thirteenth century. Then the ice slowly encroached on Iceland, Scandinavia, and the rest of Europe.

For most of Europe, the end of the good, predictable weather came suddenly. The weather had begun to deteriorate in Eastern Europe in the 1200s. By the 1300s, there was a string of wet years for all of Europe, especially summers, between 1313 and 1321. The worst was the year of 1315, when the grain "failed to ripen across Europe."[39] Catastrophic rains affected an enormous area from Ireland to Germany and north into Scandinavia. In parts of northern England, huge tracts of topsoil were washed away, leaving rocks and gullies.

The overwhelming importance of farming to the economy of that era meant that most Europeans suffered with the reduced harvests. Nearly 90 percent of the people were subsistence farm families with little left at the end of most winters except the few chickens for which they could spare grain, and enough seed to plant the next summer's crop—if they were lucky. The quantity and quality of each harvest was crucial.

Hubert H. Lamb reports extensive hunger, deaths on a great scale, and even some cannibalism.[40] The hunger was worsened by the previous century's population growth. France's population had nearly tripled, to almost 18 million.[41] England's population had risen even more sharply, from about 1.4 million to more than 5 million.

Iceland's tax records show its population dropped from more than 75,000 in 1095 to about 38,000 in the 1780s. Polar bear skins became popular as floor coverings for churches, indicating lots of bears—and lots of sea ice carrying them to Iceland's shores.[42]

Lamb reports at least three sea floods of the Dutch and German coasts between 1362 and 1446 with death tolls of more than 100,000 each.[43] Sea levels may have risen moderately during the Medieval Warming, due to the melting of glaciers, says Lamb—but in the Low Countries the storm surges were aggravated by the sinking of the earth's crust in the North Sea Basin (which is still ongoing). The real killer, of course, was the storm surge.

Overall, the Little Ice Age persisted for 550 years, featuring cooler, occasionally stormy, and generally unpredictable weather with short growing seasons more years than not.

Wet growing seasons often left the farmer with high-moisture grain that was subject to mold during storage. However, people eat moldy grain if they have nothing else. During the Little Ice Age, they often had nothing else. The ergot fungus attacked wet grain crops, especially rye, the food of the common people. As a result there was a surge in "St. Anthony's fire," or ergotism. When people ate the rye, the fungal toxin in the grain would cause convulsions, hallucinations, and outbreaks of mass hysteria, often for whole villages. In extreme cases, internal gangrene from the ergot poisoning would cause victims' limbs to fall off, and even cause death. Medieval woodcuts show St. Anthony surrounded by detached hands and feet.

Some Christians believed the horrible weather was a sign that Satan was gaining dominance over the Earth. Many blamed witches for their suffering. More than a thousand people were burned as witches between 1580 and 1620 just in Bern, Switzerland. The small town of Wiesensteig, Germany, burned sixty-three women in 1563.[44] Johann Linden, canon of a church in Treves in 1590, explained the public mood in his diary: "Everybody thought the continuous crop failure was caused by witches from devilish hate, so the whole country stood up for their eradication."[45]

Increased Cold Equals Increased Illness

Many diseases were rampant during the Little Ice Age, encouraged by persistently damp clothing, overcrowding, malnutrition, poor sanitation, and a scarcity of heating fuel. For example, epidemics of typhus, spread by lice, claimed more lives in cold winters, because malnourished people huddled together in cramped huts to share their meager fires and body warmth. Colds and flu readily turned into pneumonia.

A typical North European dwelling had one small room, a bare earth floor, no insulation, no glass in any windows and a leaky thatched roof. People sat around the fire on low stools to stay below the smoke (which was usefully curing the bacon hung in the rafters). Wood for heating was always scarce, because it required land (forest) and huge amounts of labor with primitive tools to cut and haul it. The spread of tuberculosis was favored by crowding and malnutrition—even in castles where the wealthy were well fed.

All sorts of now-forgotten disease epidemics, from typhoid and typhus to diphtheria and whooping cough, assailed communities during the Little Ice Age as they had during the cold of the Dark Ages hundreds of years earlier.

Once Again, Travel and Trade Discouraged

All the factors that favored trade and travel during the Medieval Warming were reversed during the Little Ice Age. Shipping was assailed by bigger and less-predictable storms, and the periodic intrusion of dangerous sea ice. Carts could not be dragged through the bottomless mud on the primitive roads, caused by the heavier rains and reduced sunshine to dry them out. The mountain passes were shut by massive amounts of snow and ice. After the fourteenth century, the trade fairs waned.

The Norse colonists in Greenland grew increasingly desperate. Sea ice shortened the growing seasons for farming, and often prevented the seal hunters' boats from sailing. The cattle and sheep were increasingly short of fodder. Without native timber their isolation must have loomed more and more fearsome as the ice encroached.

1550 TO 1850: THE LITTLE ICE AGE—PHASE TWO

Then came the worst, says Lamb:

> In the middle of the 16th century, a remarkably sharp change occurred. And over the next hundred and fifty years or more the evidence points to the coldest regime—though accompanied by notably great variations from year to year and from one group of a few years to the next—at any time since the last major ice age ended ten thousand years or so ago.[46]

> During the exceptionally cold winter of 1684, Fagan reports a three-mile-wide belt of ice formed along the saltwater coast of the English Channel.[47]

On several occasions between 1695 and 1728, residents of the Orkney Isles off northern Scotland were startled to see Inuits paddling kayaks off their coasts. Once, a kayaker came as far south as the River Don near Aberdeen.[48] These hunters had come south with the encroaching Arctic ice and the seals.

Of the Vikings' Greenland settlements, Lamb writes:

> The larger Osterbygd (East Settlement)[of Greenland] where there were about 225 farms, survived until about the year 1500, though in evident decline: The average stature of the grown-up men buried in the graveyard at Herjolfsnes in the fifteenth century was only 5 ft 5 inches, compared with about 5 ft 10 inches in the early period of the settlement.[49]

In 1676, artist Abraham Hondius painted hunters chasing a fox on the frozen Thames River at London.[50] The last Thames Ice Festival was held in the winter of 1813–1814.[51]

In the extreme winter of 1695, ice blocked the entire Iceland coast in January and stayed for much of the year. The cod fishery failed completely.[52] With little hay and no fish, many of the sheep and cattle would have had to be slaughtered.

A substantial amount of marginal land had been cleared and plowed in the uplands of southern Scotland—and then had to be abandoned. "Over 20 percent of the existing moorland of the Lammermuir Hills has evidence of former cultivation. It is likely that colonization occurred in the eleventh and twelfth centuries following the introduction of the moldboard plough, but reduced temperature and increased wetness from the fourteenth century onwards made some of this land sub-marginal. Several settlements were abandoned in the seventeenth century. . . . Conditions for the people were bleak and the cultivation limits retreated downslope by as much as 200 meters."[53]

Food shortages killed millions between 1690 and 1700, and there were more famines in 1725 and 1816.[54] The food crisis brought on by the severe cold and poor harvests of 1816 ranged from the United States to the Ottoman Empire in the Middle East, into parts of North Africa, and across Europe from Italy to Switzerland. Typhus epidemics followed the harvest failures, and bubonic plague reappeared.

Jean Grove found that an unprecedented number of petitions for tax and land rent relief were granted in seventeenth and eighteenth century Norway due to landslides, rockfalls, avalanches, floods, and ice movement.[55]

China had a series of severe winters between 1654 and 1676, forcing orange groves that had existed for centuries to be abandoned in Kiangsi Province. Again we find vegetation patterns among the most powerful evidence of climate change.

Oak forests were reported in Mauritania in the seventeenth century, indicating a cooler and wetter climate south of the Sahara Desert than in recent times. Lake Chad was about four meters higher than today.[56]

The average freezing dates recorded for Lake Suwa in central Japan indicate that the mid-1800s were one of the severest phases of the Little Ice Age for that area.[57]

Rev. Ezra Stiles, then president of Yale University, started keeping a daily temperature record in 1779. His readings for June 1816 certify it as the coldest June ever recorded by a Connecticut thermometer. The average was 2.5 degrees Celsius lower than the long-term mean from 1780 to 1968.

ABC News, on 7 September 2001, featured the climate histories that generations of people worldwide have kept on lakes and river freezing dates:

These are direct observations of people, five generations of people. Some were religious people, some were fur traders. They have looked out and said, "[T]he lake, the bay, the river is open today." For example, holy people of Japan's Shinto religion kept careful records at Lake Suwa, where deities from shrines on either shore were believed to have used surface ice to visit back and forth. At Lake Constance, on the border of Germany and Switzerland, congregations at two churches, one in each country, had a tradition of carrying a Madonna figure back and forth across the lake when it froze. In Canada, the shipping and fur trade meant records of river freezing were kept as far back as the early 1700s. . . . Writing in the journal *Science*, John Magnuson and colleagues say these and other collected records tell a very clear story—lakes and rivers now freeze an average of 8.7 days later than they did 150 years ago and ice cover starts breaking up 9.8 days earlier. These findings correspond to an increase of nearly 4 degrees Fahrenheit (1.8 degrees Celsius) in air temperature over the past 150 years.[58]

ABC News, of course, was trying to scare us about man-made global warming. What they actually did was affirm the long record of human observations about ongoing changes in the earth's cycling climate.

Oceanographer William H. Quinn reconstructed the fluctuations in Nile floods, from 641 to modern times, using a wide variety of historical sources. Quinn found that during the Dark Ages, from 622 to 999, the Nile floods were below normal 28 percent of the time. During the Medieval Warming, from 1000 to 1290, the Nile floods were below average only 8 percent of the time. During the Little Ice Age, the Nile floods weakened again, falling below average 35 percent of the time.[59]

Climate Change Reflected in Art and Fish

Among our most fascinating sources of climate history are the paintings done by artists through the years of the warming/cooling cycle. Hans Neuberger studied the clouds shown in more than 6,000 paintings done between 1400 and 1967.[60] His statistical analysis shows a slow increase in cloudiness between the beginning of the fifteenth and mid-sixteenth centuries. Low clouds (a strong climate cooling factor) increase after 1550; and then decline again after 1850. Artists in the eighteenth and nineteenth centuries regularly painted 50 to 75 percent cloud cover in their *summer* scenes.

THE DEBATE: WAS THERE CLIMATE CHANGE OR NOT?

Neither the Medieval Warming nor the Little Ice Age was a precise, consistent event. Only in retrospect can we say that the Little Ice Age ended about 1850. Variability remains the foremost characteristic of weather, and it takes at least a century of weather data to evaluate climate trends.

In Iceland, Hannes Finnsson compiled *Loss of Life as a Result of Dearth Years* in 1793. He wrote that hunger years were even more common before 1280. However, he concluded the main reason for improvement was better terms of trade with Denmark after that date.

The major Icelandic author on climate change was Porvaldur Thoroddsen, who, in 1916–1917, published *The Climate of Iceland through One Thousand Years*. Thoroddsen believed the island nation's climate had not changed from settlement to his day. He believed Icelanders' complaints were simply a refusal to face the reality that they lived in a harsh and highly variable climate. He wrote, "It can be said with certainty that since Iceland was settled no significant changes in the climate or weather have occurred. Sagas and annals show quite clearly that good and bad years, in the past as now, have alternated with long and short intervals. Then, as now, the sea ice came to the coasts, the glaciers were the same, the desert areas the same and the vegetation the same."[61]

A Swedish oceanographer, Otto Pettersson, in 1914 finally linked northern Europe's economic decline during the Little Ice Age to climate change—and to Iceland as an indicator. He noted that the first Iceland settlers were able to grow grain, which became impossible after the thirteenth century, and that the sea ice became worse, both in amounts and duration.[62]

After the 1920s, the strong warming surge in European and Icelandic temperatures became clear, and the debate was over.

The debate had occurred because none of the weather historians *then alive* had seen a warmer climate in their homeland. Nor were the available historic documents and legends precise enough to prove the relatively modest changes in climate. After all, the changes were only a few degrees, and the extent of sea ice had always varied erratically.

The historic evidence for the 1,500-year cycle is also incomplete and fragmented. In chapter 9, we will turn to the records provided by the Earth itself to fill in important details.

NOTES

1. George DeWan, "Long Island History," *Newsday*, 28 September 2005.

2. J. R. Dunn, "Summer's Lease," *Analog Science Fiction and Fact*, December 2000.

3. James S. Aber, "Lecture 19: Climatic History of the Holocene," <http://academic.emporia.edu/aberjame/ice/lec19/lec19.htm> July 2004).

4. John E. Oliver, *Climate and Man's Environment* (New York: Wiley, 1973), 365.

5. Oliver, *Climate and Man's Environment*, 365.

6. H. H. Lamb, *Climate, History and the Future* (London: Methuen, 1977), 156.

7. H. W. Allen notes shared with Lamb, referred to in Allen's *History of Wine* (London: Faber & Faber, 1961), 75.

8. Oliver, *Climate and Man's Environment*, 365.

9. H. H. Lamb, *Climate, History and the Modern World* (London: Routledge, 1982), 162.

10. Robert Claiborne, *Climate, Man and History* (New York: Norton, 1970), 344–47.

11. "Tree Rings Challenge History," BBC News, 8 September 2000.

12. Ibid.

13. Lamb, *Climate, History and the Modern World*, 159.

14. David Keys, *Catastrophe: An Investigation into the Origins of the Modern World* (New York: Ballantine, 1999).

15. M. Maas, *John Lydus and the Roman Past: Antiquarianism and Politics in the Age of Justinian* (London: Routledge, 1992).

16. Oliver, *Climate and Man's Environment*, 365.

17. Cyril Mango, *Byzantium, the Empire of New Rome* (New York: Scribner, 1980).

18. Lamb, *Climate, History and the Modern World*, 172.

19. R. D. Tkachuck, "The Little Ice Age," *Origins* 10 (1983): 51–65.

20. Oliver, *Climate and Man's Environment*, 365.

21. Tkachuck, "The Little Ice Age," *Origins* 10 (1983): 51–65.

22. Brian Fagan, *The Little Ice Age: How Climatic Change Made History 1300–1850* (New York: Basic Books, 2000), 17–18.

23. Lamb, *Climate, History and the Modern World*, 181.

24. Jean Gimpel, *The Cathedral Builders* (New York: Grove Press, 1961), 68.

25. Mark Overton, "English Agrarian History 1500–1850," <www.neha.nl/publications/1998/1998_04-overton.pdf>.

26. Mark Overton, "English Agrarian History 1500–1850."

27. R. Lacey and D. Danziger, *The Year 1000: What Life Was Like at the Turn of the First Millennium* (Boston: Little, Brown, 1999), 67–81.

28. Lamb, *Climate, History and the Modern World*, 207.

29. C. Ko Chen, "A Preliminary Study on the Climatic Fluctuations during the Last 5,000 Years in China," *Scientia Sinica* 16 (1973): 483–86; W. Shao Wu, and Z. Zong Ci, "Droughts and Floods in China," in *Climate and History: Studies in Past Climates and Their Impact on Man*, ed. T. M. L. Wigley et al. (London: Cambridge University Press, 1981), 271–87; and J. Zhang and T. J. Crowley, "Historical Climate Records in China and Reconstruction of Past Climates," *Journal of Climate* 2 (1989): 830–49.

30. Z. De'er, "Evidence for the Existence of the Medieval Warm Period in China," *Climatic Change* 26 (1994): 289–97.

31. Z. Feng et al., "Temporal and Spatial Variations of Climate in China during the Last 10,000 Yrs," *The Holocene* 3 (1993): 174–80.

32. Kang Chao, *Man and Land in China: An Economic Analysis* (Palo Alto, CA: Stanford University Press, 1986).

33. Y. Tagami, "Climate Change Reconstructed from Historical Data in Japan," *Proceedings of International Symposium on Global Change, International Geosphere-Biosphere Programme-IGBP*, 1993, 720–29.

34. Tagami, "Climate Change Reconstructed from Historical Data in Japan," 720–29.

35. B. Fagan, *Floods, Famines and Emperors: El Niño and the Fate of Civilizations* (New York: Basic Books, 1999), 159–77.

36. Lamb, *Climate, History and the Modern World*, 186.

37. Fagan, *Floods, Famines and Emperors*, 197.

38. Fagan, *Floods, Famines and Emperors*, 49.

39. Lamb, *Climate, History and the Modern World*, 195.

40. Lamb, *Climate, History and the Modern World*, 194.

41. Fagan, *Little Ice Age,* 31–33.

42. Lamb, *Climate, History and the Modern World*, 188.

43. Lamb, *Climate, History and the Modern World*, 191.

44. Fagan, *Little Ice Age*, 91.

45. W. Behringer, "Climate Change and Witch-Hunting: The Impact of the Little Ice Age on Mentalities," History Department, University of York, <http://www.york .ac.uk/depts/hist/staff/wmb1>.

46. Lamb, *Climate, History, and the Modern World*, 212.

47. Fagan, *Floods, Famines and Emperors*, 197.

48. Fagan, *Little Ice Age,* 116.

49. Lamb, *Climate, History and the Modern World*, 187–89.

50. Fagan, *Little Ice Age*, 48.

51. <www.thamesfestival.org/gallery/about/origins.htm>.

52. Fagan, *Little Ice Age*, 113.

53. David L. Higgitt, "A Brief Time of History," in *Geomorphological Processes and Landscape Change: Britain in the Last 1000 Years*, ed. David L. Higgitt and E. Mark Lee (Oxford: Blackwell, 2001), 17.

54. L. Ladurie and E. Ladurie, *Times of Feast, Times of Famine*, translated by B. Bray (New York: Noonday Press, 1971), 64 79.

55. J. Grove and Arthur Battagel, "Tax Records from Western Norway as an Index of Little Ice Age Environmental and Economic Deterioration," Climatic Change 5, no. 3 (December 1990): 265–82.

56. Lamb, *Climate, History and the Modern World*, 235.

57. Lamb, *Climate, History and the Modern World*, 237.

58. J. Magnuson et al., "Historical Trends in Lake and River Ice Cover in the Northern Hemisphere," *Science* 289 (2000): 1743–746.

59. William H. Quinn, "A Study of Southern Oscillation-Related Climatic Activity for A.D. 622–1990 Incorporating Nile River Flood Data," in *El Nino: Historical and Paleoclimatic Aspects of the Southern Oscillation*, ed. Henry F. Diaz and Vera Markgraf (Cambridge, UK: Cambridge University Press, 1992), 119–50.

60. Hans Neuberger, "Climate in Art," *Weather* 25, no. 2 (1970): 46–56.

61. P. Thoroddsen, *The Climate of Iceland through One Thousand Years* (Karysmannaliofn: Reykjavik, 1916–1917), Vol. 2 (1908–1922), 371.

62. Otto Pettersson, *Climate Variations in Historic and Prehistoric Times* (Goteborg, Sweden: Swenska Hydrografis Bilogy Konnor Skriften #5, 1914), 26.

Chapter Eight

The Baseless Fears

Warming Brings Famine, Drought, and Barren Soils

The Cry of Famine and the Secret Pentagon Climate Study:

"A secret report, suppressed by U.S. defense chiefs and obtained by *The Observer*, warns that major European cities will be sunk beneath rising seas as Britain is plunged into a 'Siberian' climate by 2020. Nuclear conflict, megadroughts, famine and widespread rioting will erupt across the world. . . . Deaths from war and famine run into the millions, until the planet's population is reduced by such an extent the Earth can cope. Access to water becomes a major battleground. . . . Rich areas like the U.S. and Europe would become 'virtual fortresses' to prevent millions of migrants from entering after being forced from land drowned by sea-level rise or no longer able to grow crops."[1]

Other Viewpoints:

"In a report released to Knight Ridder [by professional futurists at the Pentagon] on Monday, [is the information] that while a drastic climate change is *unlikely*, it 'would challenge national security in ways that should be considered immediately.' The 'plausible' consequences include famine in Europe and nuclear showdowns over who controls what's left of the planet, the futurists concluded. The report, commissioned by the Department of Defense's Office of Net Assessment . . . reflects the *Pentagon's policy of planning for the worst,* said author and long-time consultant Peter Schwartz. 'It's an unlikely event, and the Pentagon often thinks the unthinkable, and that's all it is, said [coauthor Doug] Randall'"[2] (emphasis added).

"Forget the hysterical claims about global warming causing floods, famine and insect-borne diseases. MIT's Richard S. Lindzen—who is on the National Academy of Sciences panel that global warming alarmists love to cite—has testified that global warming could be 'beneficial' and that carbon dioxide increase will result in 'minimal impacts.'"[3]

FIVE REASONS NOT TO FEAR
FAMINE DURING GLOBAL WARMING

First: Lessons of History

Human food production, historically, has prospered during the global warmings. We have seen in the earlier chapters the flourishing of human society during the Roman Warming and the Medieval Warming. Food production increased during previous historic warmings primarily because warming climates provided more of the things plants love: sunlight, rainfall, and longer growing seasons. During warmings there are also less of the things plants hate: late spring frosts and early fall frosts that shorten the growing season, and hailstorms that destroy fields of crops. Jorgen Olesen of the Danish Institute of Agricultural Sciences predicts that Europe's overall food production will *increase* with warming, even though some southern European regions will have crops reduced by aridity.[4]

Second: What Science Says about Food and the Modern Warming

Sunshine: Richard Willson of Columbia University (and NASA) has measured an increase in the sun's radiance of 0.05 percent per decade for the past two decades. He says the upward trend in sunlight may well have been going on longer than that. Earlier, we didn't have the precision instruments to measure that small but vital trend, but every bit of it encourages the growth of food crops.[5]

The increased temperatures of the Modern Warming may have some negative impact on crops in the southern mid-latitudes—through drier summers, for example—in places such as southern Romania, Spain, and Texas. At the same time, however, stronger sunlight will importantly increase the productivity of farmland in the northern mid-latitudes, such as Germany, Canada, and Russia. The increased food production in the very extensive northern plains would far outweigh the negative impact of slightly more arid conditions in the southern mid-latitudes.

Rainfall: Increased heat means more precipitation, as more moisture evaporates from the oceans and then falls as rain or snow. NASA says global rainfall increased 2 percent in the twentieth century compared with the tail-end of the Little Ice Age in the nineteenth century. Most of the increased moisture fell in the mid- and high-latitudes where much of the world's most productive cropland is located.[6] We can expect this to continue through the Modern Warming.

Higher CO_2 Levels: Another reason food production has tended to increase during the past 150 years is that CO_2 levels in the atmosphere have increased. The oceans give up CO_2 when they warm. The increased CO_2 not only fertilizes the plants, but enables them to use water more efficiently.

Researchers at the U.S. Department of Agriculture in 1997 grew wheat in a long plastic tunnel, varying the CO_2 levels for the grain plants from the Ice Age CO_2 level of about 200 parts per million (ppm) at one end of the tunnel to the late-1980s level of 350 ppm at the other.[7]

The findings? An extra 100 ppm of CO_2 increased the wheat production by 72 percent under well-watered conditions, and by 48 percent under semi-drought conditions. That meant an average crop yield gain of 60 percent. These results are consistent with a wide variety of CO_2 enrichment studies done in more than a dozen countries on many different crops.

Third: Farming Technology

Human food production today depends far more on farming technology than on modest climate changes. We are no more doomed to famine by the Modern Warming than we are doomed to malaria in the era of pesticides and window screens. In fact, the food abundance the world has increasingly enjoyed since the eighteenth century is primarily due to scientific and technological advances.

In 1500, Britain could feed less than one million people. By 1850, thanks to knowledge of crop rotations and improved farm machines such as the seeder and reaper, Britain fed more than 16 million people. Today, Britain has nearly 60 million people, fed mainly from its own fields.

Today's "Climate-Secure" Agriculture

Industrial nitrogen fertilizer is one of the biggest farming advances in human history. Before 1908, farmers could only maintain their soil nutrient levels by adding livestock manure or by growing more green-manure crops, such as clover. Both of those strategies require lots of land. In 1908, however, the Haber-Bosch Process began taking nitrogen from the air, which is 78 percent nitrogen. Today's farmers apply about 80 million tons of industrial nitrogen per year to maintain their soils' fertility and it doesn't cost a single acre of land.

To get 80 million tons per year of nitrogen from cattle manure, the world would require nearly eight billion additional cattle, plus five acres or so of forage land per beast. We'd thus have to eliminate half the people, clear all the forests, or use some combination of those strategies.

The Green Revolution of the 1960s tripled the crop yields across Europe and much of the Third World.

- More powerful seeds, many of them with resistance to drought and pests, made better use of the complete roster of plant nutrients (nitrogen, phosphate, and potash—plus twenty-six trace mineral elements) that soil-testing modern farmers apply to their fields.
- Irrigation assures ample moisture, often even in semiarid areas.
- Insecticides and fungicides protect the high yields of the crops both during the growing season and in storage.

In America, where high-yield farming started earlier, diaries of early settlers in Virginia's Shenandoah Valley indicate that wheat yields around 1800 were only six to seven bushels per acre. The valley's farmers today often get ten times that yield. U.S. corn yields by the 1920s had risen to about twenty-five bushels per acre. Today, the national average is more than 140 bushels, and still rising.

The same story of soaring yields and more certain harvests is playing out today over most of the world.

The African Exception

Africa is the only place in the world where per capita food production has not been increasing in recent decades. Africa's food production has been severely hampered by its ancient soils, frequent droughts, and abundant insects and diseases. There has also been a lack of adequate research for its specific soils, microclimates, and pests—and an equally damaging lack of stable governance and infrastructure on that continent.

Two recent research developments are now particularly helpful for Africa.

- Quality-protein (QP) maize, bred in Mexico's International Maize and Wheat Improvement Center, not only has higher yields but also provides more lysine and tryptophan, two amino acids that are critical for human nutrition but are lacking in most corn varieties. The QP maize is able by itself to cure many African children of malnutrition.
- Rice breeders have successfully wide-crossed the African native rice species with Asian rice varieties, to create a family of more vigorous and higher-yield new rice varieties.

More such breakthroughs for Africa's farmers can be expected if more research investments are made for and in that continent. Better roads and bridges (and better national security) would also make farm inputs less expensive and higher crop yields more marketable no matter what happens to its climate.

Sustainability

Today's high-yield agriculture is also the most sustainable in history, thanks to fertilizers, soil testing, and a twentieth-century farming system called "conservation tillage." Conservation tillage controls weeds with cover crops and chemical herbicides instead of by plowing, which invites soil erosion. The conservation farmer just discs up the top two or three inches of topsoil along with the stalks and residue from the previous crop. This process creates trillions of tiny dams that prevent wind or water erosion. The little dams also encourage water to infiltrate the root zone of the field, instead of running off into the nearest stream.

Conservation tillage cuts soil erosion by 65 to 95 percent and often doubles the soil moisture in the field. It encourages far more soil bacteria and earthworms, both because of the constant heavy supply of crop residues for them to eat and because they hate being plowed, as they are in conventional and organic farmers' fields.

Through the expanded use of conservation tillage across the United States, Canada, South America, Australia and, most recently, South Asia, hundreds of millions of acres are now sustainably more productive than ever before in history.

Another fruitful use of technology and increased sustainability will be more efficient irrigation. Primitive flood irrigation systems in the Third World use water at less than 40 percent efficiency. Center-pivot irrigation systems with trailing plastic tubes to deliver water right at the roots (minimizing evaporation) and computer-controlled to apply just the right amount of moisture to each part of the field, can approach 90 percent water efficiency. World farmers currently use about 70 percent of the fresh water humanity "uses up." As water becomes more valuable, the capital investments in high-efficiency irrigation systems will be justified.

Fourth: The Future and Biotechnology

Today's crop yields are the product of more than two hundred years of conventional trial-and-error science. But, by 2050 the world will have some seven billion affluent humans demanding the high-quality diets that only about one billion people are able to afford today. We'll also have to feed far more pets.

That means world food demand will more than double, and we're already farming half of the Earth's available land. Additional sources of higher crop and livestock yields will be needed. The world is already using plant breeding, fertilizers, irrigation, and pesticides. However, the world is only beginning to use biotechnology, our new-found understanding of Nature's genetic codes.

The first broad application of biotechnology in agriculture has been to make plants tolerant of synthetic herbicides, so we could use the environmentally safest herbicides to protect our crops more effectively from weed competition. As a result we have somewhat raised crop yields and lowered food costs in many countries.

It also happens that one of Africa's worst endemic pests is a parasitic weed called witchweed. It invades corn and sorghum plants through their roots, and the farmer never knows it's there until his crop stalk suddenly sprouts a bright red witchweed flower instead of an ear of grain.

Genetically engineered herbicide-tolerant seeds could have solved the problem. With the seed soaked in herbicide, the witchweed would have been killed as it invaded the plant roots, and the grain would have thrived. Unfortunately, activists and European governments threatened retaliation against any African government that allowed the planting of biotech-modified crops.

Now, researchers have done a genetically researched end run around the biotech Luddites. Pioneer Hi-Bred identified corn seeds with a natural tolerance for the herbicide imazopyr, and donated the germ plasm to the International Maize and Wheat Improvement Center (CIMMYT) in Mexico. CIMMYT, in turn, has bred the herbicide tolerance into African corn varieties. Corn yields are four times as high. The technology is low cost and easy for even Africa's small farms to use.

Biotechnology (BT) has also allowed plant researchers to put an ultra-safe natural insecticide found in soils into such crop plants as corn and cotton. Because of these pest-resistant plants, millions of pounds of pesticide no longer have to be sprayed into the environment or pose hazards to beneficial insects. BT cotton and corn are being planted by millions of small farmers, especially in China and India.

An important second-generation benefit of biotechnology is finding wild natural genes that can improve our crop plants. We already have one such important breakthrough. Plant explorers nearly fifty years ago found a relative of the wild potato that was resistant to the infamous late blight virus that caused the Irish potato famine in the 1840s. Unfortunately, plant breeders were never able to successfully cross that blight resistance gene into an edible, productive domestic potato. Now, three different universities have spliced the blight resistance gene into new potato varieties. This will be especially important for densely populated parts of Asia and Africa (such as Rwanda and Bangladesh) that have become more dependent on the potato's ability to produce more food per acre than any other crop.

Black Sigatoka, a new bacterial disease of bananas and plantains (important staples in much of Africa) has been spreading worldwide. Unfortunately, bananas are especially difficult to cross-breed. Fortunately, biotechnology has now produced plants resistant to Black Sigatoka, protecting the tenuous food security of tropical and subtropical Africa.

Plant researchers also believe that biotechnology is the most likely path toward drought-tolerant crops, which would be hugely important in dealing with any long-term drought problems brought by the Modern Warming. Egypt has already inserted a drought-tolerance gene from the barley plant into wheat, producing varieties that need only a single irrigation per crop instead of eight. The drought-tolerant wheat will not only take less water, but will sharply reduce salinization of the irrigated land on which it's grown. It should also be a boon on large areas of good quality land where rainfall is scarce.

Fifth: Modern Transportation

The biggest technical advantage of the modern world in dealing with weather-related famines is modern transportation. In the coming warming centuries, we will undoubtedly be able to produce enough food from the land that gets good weather in any given year to supply all of the world's food needs. Equally important, we will be able to store food safely from years of plenty to ensure food abundance in lean years; all it takes are inexpensive concrete silos and modern pesticides to keep the rats and bugs from feasting on our food reserves before we need to draw on them.

Thanks to modern transportation, we will also be able to transport the food, wherever it is grown, to wherever in the world it is needed. Think of fifteenth-century communications—ships, ox carts, and foot messengers. It could take weeks for news of a food shortage to reach from regions where people were starving to places where extra food was available. Think of the fifteenth century's cold, wet weather and dirt roads—where carts would be mired in hub-deep mud—and its plodding, undernourished oxen. It would have been very hard to get much wheat, rice, or livestock feed from the farms of one region to the cities of another. Think of the fifteenth century's little wooden ships, braving the storms to reach famine victims with food, and blown off course for days and weeks at a time while the rats and weevils in the hold gradually destroyed its vital cargo.

Today, a message of potential crop failure and impending hunger is received before the crop has failed. The news travels at the speed of light, from Earth to satellites to Earth again. Huge trucks and railcars are set next to giant grain silos, and the food moves along computer-controlled railroads or modern interstates to giant vessels waiting at the nearest port. The ships can arrive anywhere in the world in a few days and unload a million tons of supplies at an astonishingly low cost.

Japan, one of the remarkably successful societies of the twentieth century, has almost no natural resources, and very little farmland. Japan typically imports more than 40 million tons per year of food grain, feed grain, oilseeds, meat and dairy products. It keeps one month's supply in storage, one month's supply moving toward its islands on ships, and uses forward contracts to ensure future deliveries for the longer term.

In the twenty-first century, this will be possible on a larger, more consistent scale. It will become even easier and cheaper to move food from where it grows best to where people choose to live.

TWENTIETH-CENTURY FAMINES
DUE TO "FAILURES OF GOVERNMENT"

The great famines of recent history have not been caused by climate or weather problems. Primarily, they have been caused by "failures of government."

* *In the 1930s,* the Soviet Union deliberately starved millions of small farmers and their families (the total death toll was probably more than seven million) as a way for the government to take control of their land for collective farms.
* *In 1943,* as many as three million people died in what is now Bangladesh. The rice crop was smaller that year than in 1942. However, one of the major problems was that the wartime boom had given urban workers enough buying power to bid the available food away from the landless poor of the countryside. The British administration, attempting to stave off Japanese invasion, paid little attention to the rural hunger until it was too late.
* *In 1959,* China's Chairman Mao Tse-tung dragooned tens of millions of small farmers into communal farms. Rural officials then systematically overstated the communal farms' food production to curry favor with Chairman Mao. When the central government demanded one-third of the inflated harvest numbers, the farmers were left with virtually no food. During the following two years, the farmers were too weak to plant or care for crops and the famine spread to the cities as well. More than 30 million people died.
* *In 1984–1985,* Ethiopia was under the rule of a Stalinist-like military junta. After a low grain harvest, Western nations donated both food and transportation but the junta used much of the food to feed its troops. Ordinary people got food only if they left their small farm-holdings in the anti-junta highlands and took up new land of unknown sustainability far away. A million people may have starved before the junta was eventually deposed.

WHAT ABOUT FEEDING CHINA AND INDIA?

China and India have more than two billion people, and India's population is still growing. China's farmland is severely constrained by its mountains, rivers, deserts, and erratic rainfall. India's monsoons leave the country arid much of the year and its water supplies are barely adequate to achieve today's production. How can these countries be fed in the future without depriving other countries of their necessities?

Crop yields are still rising in China, in India, and around the world. If we keep investing in agricultural research, we will find new ways to continue our technological abundance. Biotech-modified cotton, corn, and soybeans are already making positive impacts on Chinese and Indian farm output. Recently, a new pest-resistant hybrid cotton variety has been genetically engineered in China that will free 600,000 hectares of cotton land for food crops.

Then, of course, there is a massive potential for farm imports to these densely populated countries. Such temperate-zone and land-rich countries as the United States, Canada, France, Argentina, and Brazil already have major capacities to expand their farm production. Such emerging agricultural powers as Poland, Romania, Turkey, and Ukraine also have major expansion potential. China and India increasingly have the ability to pay for such food imports from the profits of their nonfarm industries and exports.

We can also hope that the world's leaders have learned a crucial lesson from the high costs and low success rates of the recent "land wars." Japan seized Manchuria in the 1930s to get oil and soybean fields. Germany went to war in 1937 at least partly for "living room"—which meant farmland. Saddam Hussein invaded Kuwait for oil. All three seizures cost the invading countries heavily in deaths and military expenditures, only to be turned back by the world community of nations.

In an era of free trade, rapid communications, and ready financing, it is far more cost effective to import resources and commodities than to go to war for them.

WHAT ABOUT MEGA-DROUGHTS?

History tells us that both warming and cooling may trigger very long droughts in some regions. There is reason to believe that California, Mexico, and southern Africa all run increased drought risks as the 1,500-year climate cycle continues its solar-driven course.

When the Mayan Indians in Mexico and Central America faced this choice—during a global cooling phase—they had little knowledge of the long-term climate cycle. They had even less capacity to import food from other regions. They eventually just left their cities and faded into the jungle.

If California were faced with the likelihood of long-term drought, the state would probably consider desalinization plants for drinking water, and food imports from other regions that were getting both warmer and wetter due to climate change. In fact, the California public is already debating long-term drought strategies, with additional water reservoirs and desalinization among a wide range of choices.

Extended droughts cause damage whenever and wherever they occur. If humanity could forestall even moderate climate change, we might very well vote to do so.

However, there is no indication in the ice and sediment cores that humanity can alter the 1,500-year climate cycle. Our information from fossils, tree rings—and climate models—on where and when such droughts might occur is fragmentary and uncertain. We don't know yet where our descendants will need additional flood control or water storage—let alone the technologies they will have to provide them.

What we can say is that today's human societies are better equipped to tolerate extended droughts than any previous human societies. Fortunately, a natural warming will move along a slowly erratic course over the coming centuries, giving time for farmers, communities, and governments to adapt. These adaptations are likely to come gradually, in modest increments, as public opinion and technical feasibility determine what changes need to be made.

NOTES

1. Mark Townsend and Paul Harris, "Now the Pentagon Tells Bush: Climate Change Will Destroy Us," *The Observer*, 11 November 2004.

2. Seth Borenstein, "U.S.: Climate Change Could Cause Global Woe," Knight-Ridder-Tribune News Service, 24 February 2004.

3. Debra Saunders, "More Hot Air on Global Warming," *San Francisco Chronicle*, 2 November 2001.

4. Robert Uhlig, "Feast and Famine in Europe as Global Warming Scorches Farms," *Daily Telegraph* (London), 21 August 2003.

5. R. Willson and A. V. Mordvinov, "Secular Total Solar Irradiance Trends during Solar Cycles 21–23," *Geophysical Research Letters* 30, no. 5 (2003): 1199.

6. A. Dai, I. Y. Fund, and A. Del Genio, "Surface Observed Global Land Precipitation Variation, 1900–1988," *Journal of Climate* 10 (December 1997): 2943–962.

7. H. S. Mayeaux et al., "Yield of Wheat across a Subambient Carbon Dioxide Gradient," *Global Change Biology* 3 (1997): 269–78.

Chapter Nine

The Earth Tells Its Own Story
of Past Climate Cycles

Global Warming Advocates Say:

"The idea that there's no need to worry about human-induced global warming because the world's climate in medieval time was at least as warm as today's is flawed, according to a recent analysis. There's not enough evidence to conclude that the Medieval Warm Period was global."[1]

Informed Skeptics Say:

"Researchers have gathered comprehensive information about past climate change from proxies such as tree rings, pollen, coral, glaciers, boreholes, and sea sediments sampled worldwide. According to the reconstructed records, people in many parts of the world experienced a relative warmth early in the millennium, called the Little Optimum, and a cool period a few centuries later labeled the Little Ice Age. . . . The Little Optimum and Little Ice Age were real. They were also widespread over the globe. The 20th century is not the least bit climatically unusual. So, why the recent media hysteria that the 20th century is the warmest of the last 1,000 years?"[2]

Analysis of a 160-meter ice core removed from Wyoming's Upper Fremont Glacier: "The termination of the Little Ice Age was abrupt with a major climatic shift to warmer temperatures around 1845 A.D. . . . [A] conservative estimate for the time taken to complete the Little Ice Age climatic shift to present-day climate is about 10 years."[3]

TEMPERATURE PROXIES

Many items can be tested and the resulting data used as temperature proxies. Such items include ice cores, seabed sediments, boreholes, tree rings, tree lines, stalagmites, and pollen. An overview follows.

Ice cores were the first breakthrough in tracing past climate cycles and the long ice cores have all been brought up since 1980. The ice cores testify with remarkable consistency to a constantly changing climate on our planet, defined mainly by two massively persistent cycles: the 100,000 year cycle of ice ages and interglacials, and the 1,500-year cycle of moderate warmings and coolings that runs through at least the last one million years.

The first long ice cores came from the Greenland Ice Sheet and were analyzed by Denmark's Willi Dansgaard and Switzerland's Hans Oeschger in 1983. They went back 250,000 years.[4] A few years later, the Vostok Glacier ice core was brought up at the other end of the world in Antarctica. It went back 400,000 years and it, too, showed the 1,500-year cycle.[5] Thus, the ice cores from both ends of the Earth tell us of the same 1,500-year cycle extending far back into prehistory.

Close to the surface, the ice cores' layers can be counted visually. For longer cores, researchers use such techniques as measuring variations in hydrogen peroxide. Since the peroxide is created by ultraviolet sunlight, and there's virtually no sunlight in the Antarctic winter, the hydrogen peroxide levels for summer are five times higher than for winter and easily identifiable.

To read temperature change, mass spectrometers often measure the depletion of oxygen-16 isotopes in ice, sediments, or wood compared to the less-abundant oxygen-18 isotopes. Variations in deuterium (heavy hydrogen) and carbon-14 are also used as proxies.

Seabed sediment cores are also analyzed for the remains of tiny phytoplankton and zooplankton species. Their types and abundance indicate temperatures, and their tiny fossils can be carbon dated.

Diatoms in a sediment core from Cameroon's Lake Ossa in West Africa show that the climate there oscillates with the 1,500-year cycle in the northern and southern movements of the Intertropical Convergence Zone.[6] V. Francis Nguetsop of the French National Museum of Natural History says that a southward shift of the ITCZ was marked by low precipitation in the northern subtropics (i.e., Nigeria and Ghana) and high precipitation in the subequatorial zone (i.e., Zaire and Tanzania).[7]

In East Africa, Belgium's Dirk Verschuren built a 1,100-year rainfall-drought history for Kenya's Lake Naivasha, based on (1) sediments, (2) fossil diatoms, and (3) midge species and numbers. "In tropical Africa," Verschuren says, "the data indicate that, over the past millennium, equatorial East Africa has alternated between contrasting climate conditions, with significantly drier climate than today during the 'Medieval Warm Period' (A.D. 1000–1270) and a relatively wet climate during the 'Little Ice Age' (A.D. 1270–1850) that was interrupted by three prolonged dry episodes."[8]

Near the Mayan cities of Central America, titanium levels in seabed cores from the Venezuelan coast testify to a prolonged drought during the cold Dark Ages that may have caused the collapse of the entire Mayan culture.[9]

Boreholes give accurate temperature histories for about 1,000 years into the past because rock transmits past surface temperatures downward. The University of Michigan's Shaopeng Huang led a study of 6,000 boreholes in 1997 from all continents. The results clearly showed that temperatures during the Medieval Warming were warmer than those of today and that during the Little Ice Age they fell 0.2 to 0.7 degrees Celsius below present.[10]

Tree rings can obviously be counted to date the time of an event and their summertime width is greater under good growing conditions (warmth, rainfall) than during poor growing seasons (cold, dry). They are limited by the distance back in time researchers can find live trees, dead trees, buried wood, or even structural wood from an earlier time which can be accurately dated to its growth period.

In mountainous northwestern Pakistan, more than 200,000 tree-ring measurements were assembled from 384 long-lived trees that grew on more than twenty individual sites. The 1,300-year temperature proxy shows the warmest decades occurred between 800 and 1000, and the coldest periods between 1500 and 1700.[11]

Mountain tree line elevations are another sensitive and highly accurate proxy for temperature change. A number of studies of European tree lines testify to the fact that tree lines, farming, and villages moved upslope during the Medieval Warming and back down with the Little Ice Age.

A recent study of tree line dynamics in Western Siberia showed that advances in the tree lines during the warmer weather of the twentieth century were "part of a long-term *reforestation* of tundra environments." Two Swiss scientists, Jan Esper and Fritz-Hans Schweingruber, note that "stumps and logs of *Larix sibirica* can be preserved for hundreds of years" and that "above the tree line in the Polar Urals such relict material from large, upright trees were sampled and dated, confirming the existence, around A.D. 1000, of a forest tree line 30 meters above the late 20th century limit." They also note, "this previous forest limit receded around 1350, perhaps caused by a general cooling trend." Thus, the Siberian tree lines testify to the Medieval Warming and the Little Ice Age well outside of Europe.[12]

Lisa J. Graumlich of Montana State University combined both tree rings and tree lines to assess past climate changes in California's Sierra Nevada. The trees in the mountains' upper tree lines are preserved in place, living and dead, for up to 3,000 years. Graumlich says:

> A relatively dense forest grew above the current tree line from the beginning of our records to around 100 B.C., and again from A.D. 400 to 1000, when temperatures were warm. Abundance of trees and elevation of tree line declined very rapidly from A.D. 1000 to 1400, the period of severe, multidecadal droughts. Tree lines declined more slowly from 1500 to 1900 under the cool temperatures of the Little Ice Age, reaching current elevations around 1900.[13]

Graumlich's tree evidence confirms both of the last two 1,500-year cycles: the Roman Warming/Dark Ages climate cycle and the Medieval Warming/ Little Ice Age. Severe drought, which has been documented in California during the latter part of the Medieval Warming, obscured the timing of the shift from the Medieval Warming to the Little Ice Age. However, both events were clearly evident.

Cave stalagmite cores confirm the global nature of the 1,500-year cycle found in ice cores, seabed sediments, and trees. Their carbon and oxygen isotopes, and their trace element content, vary with temperature. Moreover, the stalagmites go back further in time than the tree evidence. Cave stalagmites have been found in Ireland, Germany, Oman, and South Africa whose layers all show the Little Ice Age, the Medieval Warming, the Dark Ages, and the Roman Warming.[14] A number of the stalagmites also show the unnamed cold period that preceded the Roman Warming.

In southern Ontario, pollen shows that the warmth-loving beech trees of the Medieval Warming gradually gave way to cold tolerant oaks as the Little Ice Age came on—and then the forest became dominated by pine trees. The oak trees have been making a comeback in Ontario since 1850 and the beech trees can be expected to resurge as the Modern Warming continues in the centuries ahead.[15]

Remains of prehistoric villages in Argentina were analyzed by Marcela A. Cioccale of the National University of Cordoba to determine where Argentina's native peoples lived over the past 1,400 years. Using carbon-14 dating, she found that the inhabitants clustered in the lower valleys during the Dark Ages period, and then moved higher up the slopes as the Medieval Warming brought "a marked increase of environmental suitability, under a relatively homogeneous climate." Habitation moved up as high as 4,300 meters in the Central Peruvian Andes around 1000 as the Medieval Warming not only raised temperatures but created more stable conditions for farming. After 1320, people migrated back down the slopes as the colder, less stable climate of the Little Ice Age set in.[16]

Yang Bao of the Chinese Academy of Sciences reconstructed China's temperature history for the last 2,000 years from ice cores, lake sediments, peat bogs, tree rings, and the historic documents that date back farther in China than in any other country. He found China had its highest temperature during the second and third centuries, toward the end of the Roman Warming, China's climate was also warm from 800 to 1400, cold from 1400 to 1920, and then began to warm again after 1920.[17] (See figure 9.1.)

Researchers have used hundreds of different proxies to identify past temperature changes. Two of the more interesting ones are from Greenland:

• During the Little Ice Age, birds became scarce on the island of Raffles O, just east of Greenland. In the last one hundred years, as the region has warmed, the birds have returned in large numbers. This is confirmed by "an increase in organic matter in the lake sediment and by bird observations." Based on the chemistry of the sediments, however, the bird numbers are still not as large as they were during the Medieval Warming.[18]
• Henry Fricke of the University of Michigan tested the tooth enamel of dead Vikings for the ratios of oxygen-18 to oxygen-16. Comparing the tooth enamel of skeletons buried in 1100 with those buried in 1400, he documented a 1.5 degree Celsius drop in temperatures.[19]

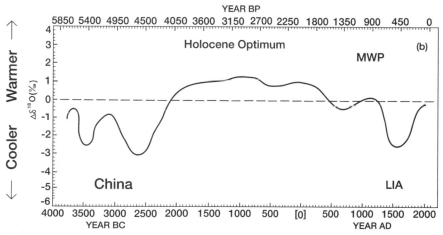

Figure 9.1.   2000 Years of Chinese Temperature History. Roman Warming temperatures were very high about A.D. 200.

Source: Y. T. Hong et al., "Response of Climate to Solar Forcing Recorded in a 6,000-Year Time-Series of Chinese Peat Cellulose," *The Holocene* 10 (2000): 1–7.

MEDIEVAL WARMING AND
LITTLE ICE AGE—SUMMARY OF THE EVIDENCE

Willie Soon and Sallie Baliunas of the Harvard-Smithsonian Center for Astrophysics have demonstrated the dominance of the climate change in the
Earth's history. They conducted a metanalysis of the research on physical
evidence of the last climate cycle, the Medieval Warming/Little Ice Age.
They found 112 studies containing information about the medieval warm
period. Of these, 92 percent showed evidence for the Medieval Warming.
Only two showed no evidence of warming.[20] The studies *confirming* the
warm period covered the land areas of Greenland; Europe; Russia; the U.S.
Corn Belt, Central Plains and Southwest; much of China and Japan; southern Africa; Argentina, Chile and Peru in South America; Australia and
Antarctica. At sea, the warming was found in the North Atlantic; off the
U.S. Mid-Atlantic Coast; off the Atlantic coast of West Africa; in the south
Atlantic near Antarctica; in the Indian Ocean, both central and southern;
and in the Central and Western Pacific Ocean. In the Southern Hemisphere,
twenty-one of twenty-two studies (95 percent) showed evidence of the Medieval Warming.

Soon and Baliunas found 124 research studies addressing the existence of
the Little Ice Age, and 98 percent contained evidence confirming that cold period. The regions where it was confirmed include the land areas of Greenland;
Europe; Russia; Asia's Himalayas; much of China and Japan; northwestern
and southern Africa; Argentina, Peru and Chile in South America; Australia,
New Zealand, and Antarctica; and, in the United States, the Corn Belt, Central Plains, and the Southwest. In the oceans, the Little Ice Age was identified
in the North Atlantic; in the Mediterranean; in the Caribbean Sea; off the West
African coast in the central Atlantic; in the South Atlantic; in the central and
southern Indian Ocean; and in the Tasman Sea south of Australia. In the
Southern Hemisphere, twenty-six of twenty-eight studies (93 percent)
showed evidence of the Little Ice Age. That may not be a large number of
studies compared to the research on Northern Europe, but their evidence is far
from trivial and their near-unanimity is impressive.

Soon and Baliunas found 102 studies containing information on whether
the twentieth century was the warmest (or most unusually warm) on record.
Seventy-eight percent found earlier periods, lasting at least fifty years that
were warmer than any period in the twentieth century. Only three studies said
the twentieth century was the warmest. Four studies rated the *early* part of the
twentieth century, before humans released much CO_2 into the atmosphere, as
the warmest or most unusually warm.

ARE YOU CONVINCED YET?

For some readers, this global sampling of the physical evidence, plus our assurance that there is a plethora of documentation, may be enough to convince. However, for the fascinated or the skeptical, the rest of this chapter is devoted to additional evidence of the 1,500-year climate cycle. Some readers may want to look at particular regions, such as the Southern Hemisphere or Africa. For the determined skeptics, let us point out that there are hundreds of other, similar research studies documenting still more physical evidence of past climate change on Earth not included here for sheer lack of space.

THE TASMANIAN TREE RING CONTROVERSY

A study of Tasmanian tree rings published in 2000 illustrates the depths of scientific disagreement which have emerged during the global warming debate—and the vital importance of looking carefully at the Earth's physical evidence of past climate change. The UN-IPCC's 2001 report claims that the Southern Ocean region may not have experienced the Little Ice Age:

> [R]ecords derived from [cave] stalagmites and stalactites and glacier evidence from the Southern Alps of New Zealand suggest cold conditions during the mid-17th and mid-19th centuries (Salinger, 1995[21]). [Tree ring] evidence from nearby Tasmania (Cook et al., 2000) shows no evidence of unusual coldness at these times. Differences in the seasons most represented by this proxy information prevent a more direct comparison.[22]

There was a big media splash when Edward Cook of Columbia University's Lamont-Doherty Earth Observatory and his team published "Climatic Change in Tasmania Inferred from a 1089-Year Tree-Ring Chronology of Huon Pine" in *Science* in 1991. The authors said, "A climatically sensitive huon pine tree-ring chronology from western Tasmania allows inferences about Austral summer temperature change since A.D. 900. Since 1965, huon pine growth has been unusually rapid for trees that are in many cases over 700 years old. This growth increase correlates well with recent anomalous warming in Tasmania on the basis of instrumental records and supports claims that a climatic change, perhaps influenced by greenhouse gases, is in progress."[23]

No wonder the IPCC's man-made-warming campaigners wanted to call attention to Cook's paper again, a decade after its publication. But that decade had given other researchers time to examine the Cook study—and find its damning flaws.

The Cook team violated good scientific practice by matching tree rings from the cool, wet climate of western Tasmania with temperature records from urban heat islands on the drier eastern side of the island. By doing so, they made it look as though global warming's recent higher temperatures had caused a strong growth spurt in the huon pine.

In fact, the temperature increase in western Tasmania during the twentieth century has been only about one-third of the 1.5 degrees Celsius credited to Tasmania in the Cook study. An average west-island temperature station reading of about 13°C early in the century had risen to only about 13.5°C in the latter part of the century.[24]

In sharp contrast, the temperature records selected by the Cook team to represent Tasmania came from the eastern, drier side of the island—in urban heat islands:

- Hobart is the biggest city in Tasmania (population 135,000). Manuel Nunez of the University of Tasmania documented a 5 degree Celsius heat island effect for the city in 1979. (And that's compared to the relatively warm and dry eastern Tasmanian countryside around it, not to the wet, wind-swept hills of the island's western coast.)
- Launceton, the second record selected by the Cook team, is the second-largest city in Tasmania, with a population of 75,000. It is also akin to Los Angeles and Mexico City, being located in an inversion bowl. Like them, it is surrounded by hills and mountains that trap its heat. Winter frosts settle frequently on the surrounding countryside but not on Launceton.
- The third temperature record in the Cook paper is the worst of all: Low Head Lighthouse. The lighthouse is at the mouth of the Tamar River on the island's North Coast, surrounded by a big aluminum smelter, an oil-fired power plant, a heavy industrial complex and an urban center with 7,000 people. Offshore breezes at night carry air from the industrial complex past the lighthouse. In addition to the industrial activity at Low Head, the lighthouse's temperature record showed a 1 degree Celsius increase in temperature since 1960, which was unique in Tasmanian climate records. Investigation found that shrubbery had grown up near the measuring box, screening it from the prevailing northwesterly winds, and turning the "official" measuring box into a sun-trap.[25]

Could Cook and his team have searched the Tasmanian climate records for the three stations with the highest temperature increases and matched them up, willy-nilly, with their west-coast tree ring data? The Cook paper obviously does not demonstrate radically higher Tasmanian temperatures, as the IPCC wanted us to believe. The temperatures in western Tasmania certainly did not rise high enough to match the heat-island increases in Hobart and Launceton.

Now for the second big flaw in the Cook et al. study: What could have caused the surge in growth of the huon pine if the temperatures were increasing only moderately? Almost certainly, here is another dramatic case of increased CO_2 in the atmosphere sharply boosting growth rates of already-mature trees. To the huon pine, the additional CO_2 in their air must have seemed like an oxygen tank to a struggling marathon runner.

The Cook authors made no mention in their paper of the "fertilizer" effect of CO_2 enrichment on trees in general since 1940—or on the huon pine in their study. It is hardly possible that the Cook team did not know that additional CO_2 in the air stimulates tree growth. After all, they're tree-growth experts and CO_2 enrichment has been a worldwide phenomenon for more than half a century.

How did the Cook paper pass its peer review to get published in *Science*? Why did the UN's panel on climate change give this badly conceived and deeply flawed study equal weight with the much more extensive and fully-peer-reviewed evidence of the Little Ice Age from more than one hundred nearby New Zealand glaciers and cave stalagmites?

The Salinger assessment cited by the IPCC notes: "From [the South Island of New Zealand's] mountain glacier records, cooler periods of climate occurred in the 11th century, early 12th century, mid 13th century, early 15th century, early 16th century, 17th and 18th centuries, and the mid-19th century."[26] That suggests there was no pattern to the cooling.

The proxy record in New Zealand, however, clearly shows the Medieval Warming, the Little Ice Age, and the Modern Warming. Oxygen and carbon isotope ratios from the New Zealand stalagmite, for example, show a sharp drop in temperature in the 1400s and cooler temperatures until the mid-1800s, after which the Modern Warming is seen. In fact, A. T. Wilson and Chris Hendy's research team remarked on the similarity of the New Zealand temperature changes to those of England in the same period.[27]

The comparison of the data in both quantity and quality makes *Science*'s peer review look inadequate and the IPCC look political. Selecting the Cook paper is at best a case of scientific bias and may well be worse.

FROM GLOBAL NORTH TO SOUTH: MORE PHYSICAL EVIDENCE OF PAST CLIMATE CYCLES

The World of Glaciers

Dismay When They Retreat:

"The glaciers of Mount Kilimanjaro in east Africa and the Andes of Peru are melting so fast that they could disappear within 10 to 20 years. The news follows other warnings that the Arctic ice field is both shrinking in area and thinning in depth. A glacier in Antarctica has also retreated dramatically in the past decade. Now, according to Lonnie Thompson of Ohio State University, the Quelccaya glacier in Peru has retreated 32 times faster in the past two years than in the 20 years from 1963 to 1983. Kilimanjaro's ice fields have retreated by at least 80 percent since 1912. The icecap of Mount Kenya has shrunk by 40 percent since 1963. In 1972, in Venezuela, there were six glaciers; now there are only two. They too will melt within a decade. . . . 'These glaciers are very much like the canaries once used in coal mines,' said Professor Thompson."[28]

And, Dismay When They Advance:

"The year was 1645, and the glaciers in the Alps were on the move. In Chamonix at the foot of Mont Blanc, people watched in fear as the Mer de Glace (Sea of Ice) glacier advanced. In earlier years, they had seen the slowly flowing ice engulf farms and crush entire villages. They turned to the Bishop of Geneva for help, and . . . at the ice front he performed a rite of exorcism. Little by little, the glacier receded. . . . Similar dramas unfolded throughout the Alps and Scandinavia during the late 1600s and early 1700s as many glaciers grew farther down mountain slopes and valleys than they had in thousands of years. Sea ice choked much of the North Atlantic. . . . [I]n China, severe winters in Jiang-Xi province killed the last of the orange groves that had thrived there for centuries."[29]

But retreating and advancing is what a glacier does. Glaciers are an ever-present, ever-changing part of our Earth's environment, growing and advancing during coolings, shrinking during warmings, and sometimes even disappearing. Most, of course, are in the Earth's Polar Regions, but others are scattered in diverse parts of the globe, wherever the elevations are high enough to turn rain into ice and temperatures seldom get above the melting point.

Humans worry (perhaps excessively) when glaciers shrink but we suffer dreadfully when they truly advance. Picture a mile-thick glacier encasing the middle of Chicago—something primitive human hunters might actually have seen during the 90,000 years of the last ice age.

Glacier advances and retreats can be dated through carbon-14 from lichens and organic material in the debris at their farthest advances. Historical records tell us of glaciers retreating during the Medieval Warming and advancing again like huge bulldozers during the Little Ice Age. The wood and other organic debris tell us their most recent advances occurred in the Little Ice Age—but researchers say the piles of rocky rubble that mark the glaciers' furthest advance often contain datable material from more than one advance!

Johann Oerlemans of Utrecht University in the Netherlands was cited in the 2001 Third Assessment Report of the IPCC.[30] His graph pointed out that the major glaciers of the world all started to shrink around 1850, but half of them stopped shrinking around 1940. Many have started to grow since 1940.

The immediate future of the glaciers is not certain, according to Oerlemans. Glaciers' response to climate change varies importantly over time with their shape, location, and elevation, and with precipitation, humidity, and cloudiness. He and his team modeled the responses of twelve different valley glaciers (not including the big subpolar glaciers) to six different climate scenarios. They found that in a rapid warming of 0.04 degrees Celsius per year, few valley glaciers would survive until 2100—if there were no change in precipitation. On the other hand, a slower warming of 0.01 degree Celsius per year—with a very likely 10 percent increase in rain and snowfall—would reduce the glaciers by only 10 to 20 percent of their 1990 volume by 2100.[31]

The one thing we can say for certain about glaciers is that the next Little Ice Age is likely to expand most of them and the next major Ice Age will see massive amounts of ice reigning again over the global landscape.

THE ARCTIC AND ITS GLACIERS

The eighteen Arctic glaciers with the longest observation histories were examined in 1997.[32] More than 80 percent of them had lost mass since the end of the Little Ice Age. Surprisingly, however, *there's no evidence the Arctic glaciers have shrunk faster during the CO_2-enriched twentieth century*. In fact, the researchers say the glaciers are losing *less* mass per year as time goes by.

Arctic glacial advances and retreats over the past 7,000 years were gauged by carbon dating the outer growth rings of trees killed when the glaciers advanced, and estimating the lichen ages in the moraines left behind when the glaciers retreated. The University of Colorado's Parker Calkin says the glaciers retreated during the Medieval warming for "at least a few centuries before 1200," and then advanced three times during the Little Ice Age: the early fifteenth century, the middle seventeenth century—and the last half of the nineteenth century.[33]

On the Russian Island of Novaya Zemlya in the Arctic Ocean, the glaciers retreated rapidly before 1920 but the retreat then slowed.[34] After 1950, more than half of the glaciers stopped retreating and many tidewater glaciers began to advance. Why? The island's temperatures in the last four decades have been lower than the previous forty years—in both the winter and the summer. That's counter to the Arctic warming predicted for the twenty-first century by climate models, particularly for the winter season.

The physical evidence tells us that the Arctic glaciers have indeed expanded as the earth's temperatures have declined and contracted as the temperatures warmed, clearly showing both the Medieval Warming, the Little Ice Age, and in some cases earlier temperature cycles. The Arctic glaciers also testify that their region has not recently been in a warming trend. Jean Grove, one of the top authorities on that period of climate history, recently did a review of the scientific literature to date the beginning of the Little Ice Age. She concluded it began "before the early 14th century" in regions surrounding the North Atlantic, and that "field evidence clearly shows that glaciers *on all continents* expanded and fluctuated about forward positions during recent centuries"[35] (emphasis added).

During the warmer temperatures since 1850, many European glaciers have retreated again. That's strong evidence of a global warming trend but it doesn't tell us whether the warming is natural or man-made. Moreover, Exeter University's Chris J. Caseldine says he sees "no obvious common or global trend of increasing glacier melt." He says Scandinavian glaciers have actually been growing and that glaciers in Russia's Caucasus Mountains are "close to equilibrium."[36] And, there's been a significant recent expansion of the ice mass in northwestern Sweden's famous Storglaciaren over the past thirty to forty years.[37] The University of Manchester's R. J. Braithwaite and South Korea's Yunfen Zhang concluded that it increased its mass *during the 1990s*.[38]

Moving South

In Northern Iceland: Caseldine found the maxima of four glaciers occurring in 1868, 1885, 1898, and 1917.[39] That clearly shows the gradual end of the Little Ice Age. Two of the glaciers kept retreating all the way to 1985, the time of Caseldine's northern Iceland study. The other two glaciers not only stopped retreating but have periodically re-advanced when temperatures dropped below about 8°C—which has happened several times during recent decades.

In Southern Iceland: Southern Iceland's Solheimajokull glacier has expanded and retreated repeatedly during the past three hundred years, and is currently halfway between its maximum and minimum for that period. New Zealand's A. N. Mackintosh says, "[T]he recent advance (1970–1995) resulted from a combination of cooling and enhancement of precipitation."[40] Colder and wetter means more ice.

From Tibet: An ice core from the Dasuopu glacier in the central Himalayan Mountains shows low temperatures in the first century, followed by a warming that peaked during 730–950. Then the glacier slid into a persistent cold period lasting until 1850.[41] It is again retreating during the Modern Warming.

In Italy: The Ghiacciaio del Calderone, in the Italian Apennines, is the southernmost glacier in Europe. Historic records indicate that it has currently lost about half the mass it held in 1794, the earliest record of its surface area. Maurizio D'Orefice of the Italian Geological Service says the Calderone lost ice volume very slowly from 1794 to 1884 and then melted more rapidly until 1990.[42]

In North America: The glacial moraines around Alaska's Prince William Sound show that the ice advanced during the Little Ice Age[43] as do the tree rings of glacial moraines in the Sierra Nevada Mountains.[44]

In South America: Glaciers on the eastern side of the Andes in Peru, Chile, and Argentina's Patagonian tip all advanced during the Little Ice Age.[45] (The eastern side is protected from the vagaries of the Pacific Decadal Oscillation.) The Peruvian glaciers were most extensive in the seventeenth century, those in Patagonia (surrounded by heat-retaining oceans) during the nineteenth century.

African Glaciers: Even tropical glaciers show the Little Ice Age and the Modern Warming. We have less data on the tropical glaciers than on the rest of the world's ice sheets. However, a study by the University of Innsbruck's Georg Kaser shows the glaciers of South America, Africa, and New Guinea all reached their greatest extents during the Little Ice Age. They've been receding "since the second half of the 19th century," just as the end of the Little Ice Age would lead us to expect. The 1930s and 1940s brought a marked loss of the tropical glaciers' ice masses. Around 1970, the melting generally slowed and some glaciers even advanced. The 1990s again brought "marked glacier recession on all tropical mountains under observation."[46] Here, too, glaciers are lagging indicators of temperature shifts.

Environmentalists have worried particularly about the glacier on Africa's Mount Kilimanjaro, because it has shrunk from more than 12 square kilometers of ice to 2.2 square kilometers since 1880. Dr. Kaser recently led a team of twenty international scientists to examine the declining ice mass.[47] They concluded that the Kilimanjaro glacier has shrunk because the climate around the mountain became much dryer. The year 1880 was, of course, long before any significant human emissions of CO_2.

That's not the only curious fact about the Kilimanjaro glacier. Between 1953 and 1976—a period of global cooling—a full 21 percent of the glacier's maximum area disappeared. After the global cooling trend became a *warming* trend (about 1979) the Kilimanjaro glacier *slowed its retreat*. (The satellites say the region around the mountain cooled during that period, despite the global warming trend.) "[Higher] air temperatures have not contributed to the recession process so far," says the Kaser team.[48]

In New Zealand: Moraines of more than 130 glaciers in New Zealand have been dated, mostly in the Westland National Park and the Hooker Range. They show three particular periods of glacial advance during the Little Ice Age, with the farthest advances in 1620, 1780, and 1830.[49] The Mueller Glacier on Mt. Cook, and the Tasman Glacier also reached their greatest extent during the Little Ice Age.[50]

In the South Shetland Islands: Glaciers just north of Antarctica also advanced during the Little Ice Age, based on the age of lichens[51] and analysis of lake sediments.[52]

The Antarctic Glaciers: On the Scott Coast of Antarctica, the Wilson Piedmont Glacier advanced at approximately the same time as the main phase of the Little Ice Age. The advance was plotted using a technique somewhat similar to carbon dating the lichens on land-based glacial moraines: The researchers used carbon-14 dating to analyze the organic material in the glacier's raised beaches. (They also used aerial photos and direct observations taken since the late 1950s.)[53]

Note that more than 130 Southern Hemisphere glaciers advanced during the Little Ice Age, most of them in well-known locations in New Zealand! Why did the IPCC ignore this massive confirmation of global cooling in the Southern Hemisphere, preferring instead to cite a seriously flawed study of huon pine in Tasmania?

Other Temperature Proxies

Greenland: This is an outstanding place to take ice cores since it offers high rates of ice accumulation, a simple ice flow pattern, lots of ice thickness (for a more detailed record), and a significant location in the weather patterns of the North Atlantic.

Dorthe Dahl-Jensen and her research team from the University of Copenhagen reconstructed the temperature history of Greenland from two Greenland ice sheet boreholes. "The record implies that the medieval period around 1000 A.D. was 1 [degree C] warmer than the present in Greenland. Two especially cold periods, at 1550 and 1850 A.D. are observed during the Little Ice Age, with temperature –0.5 and –0.7 [degrees C] below the present. After the LIA, temperatures reached a maximum around 1930; temperatures have decreased during the last decades."[54] (Her team's study ended with 1995.)

Sediment cores from a fjord in East Greenland confirm an initial cooling around 1300, with very severe and variable climatic conditions from 1630 to 1900.[55]

Off Alaska: Old Dominion University's Dennis Darby led a team analysis of sediments from the continental shelf off Alaska.[56] The number and species of dinocysts (tiny "cocoons" left behind by one-celled organisms) gives evidence of sea surface temperatures and sea-ice cover. The most surprising result of their study was the large variation in Arctic temperatures shown by the proxies—6 degrees Celsius over the last 8,000 years, a greater range than on the Greenland Ice Sheet.

In Central Alaska: Evidence of warmer past temperatures was also found in central Alaska, including the expansion of forest ranges and the absence of permafrost during the prior interglacial era.[57] Estimated summer temperatures were at least 1 to 2 degrees Celsius higher than now and in some locations may have been as much as 5 degrees Celsius warmer.

Much of Alaska has been warming in recent decades. However, the northern Pacific Ocean is characterized by large, sudden climatic shifts (essentially the Pacific Decadal Oscillation with thirty-year phases).[58] Researchers have identified eleven major shifts in this region since 1650. Alaska has been running counter to the cooler temperature pattern in the rest of the Arctic, with the help of the Pacific currents. However, the 1976–1977 PDO shift is probably responsible for the vast majority of Alaska's recent temperature rise.

Northern Quebec: The climatic history was reconstructed over the past 4,000 years using "ice wedges," large soil deformations which start as vertical cracks in the soil, fill with water in the summer, and freeze into their characteristic flat-topped wedge shapes as the soil contracts—much as mud dries in the sun.[59] They can build up in size for centuries. The wedges indicate that the region was severely cold between 1500 and 1900—the Little Ice Age. Measurements show colder conditions again during the last fifty years.

The ice wedge study was confirmed in Northern Quebec by tree rings and growth sequences from more than three hundred spruce tree skeletons buried in a peat land near the tree line.[60] The trees showed colder weather from 760 to 860, a warming from 860 to 1000, and severe cold from 1025 to 1400.

Moving South into Central Europe

Lund University's Bjorn Berglund did one of the broadest assessments of European climate proxies, including solar exposure, glacier activity, lake and sea levels, bog growth, tree lines, and tree growth. Berglund documented the Dark Ages cooling in various tree ring studies, as well as with algae microfossils in the Norwegian Sea. As we mentioned earlier, he found a major retreat from farming that started about 500, extending over "large areas of central Europe and Scandinavia."[61]

Berglund says the Dark Ages were followed by a "boom period," for farming from 700 to 1200, correlating with the Medieval Warming. "The climate was warm and dry, with high tree-lines, glacier retreat and reduced lake-catchment erosion."[62] (That means there were fewer big storms.) After 1200, the European climate became cooler and wetter with the beginnings of the Little Ice Age.

Scandinavia: A 1,400-year tree ring study of the region's summer temperatures, led by Keith Briffa of Britain's University of East Anglia in 1990,[63] showed little evidence of the Medieval Warming or Little Ice Age.

In 1992, however, Briffa and several of the same coauthors published another report in *Climate Dynamics*, noting that "our previously published reconstruction was limited in its ability to represent long-timescale temperature change because of the method used to standardize the original tree-ring data. Here we employ an alternative standardization technique which enables us to capture temperature change on longer timescales."[64]

This second report found a cool period from 500 to 700, with 660 an especially cold year. Then it showed generally warm periods from 720 to 1360 (the Medieval Warming) with "peaks of warmth" in the tenth, eleventh, twelfth, and early fifteenth centuries—up to 1430.

In Ireland: Frank McDermott, University of Dublin, analyzed a cave stalagmite for oxygen-18 isotopes.[65] The variations were broadly consistent with a medieval warm period 800 to 1200 years ago, and a two-stage Little Ice Age that matches the profiles in the Greenland Ice Sheet. The McDermott team also found the profiles of the Roman Warming and the cold Dark Ages.

In Germany: Stalagmites yield proxy records that go back more than 17,000 years, showing the Little Ice Age, the Medieval Warming, the Roman Warming, and the unnamed cold period just before the Roman era.[66] The authors specifically note that their stalagmite records resemble those of the McDermott stalagmite in Ireland.

In Switzerland: Carbon and oxygen isotopes in a sediment core from Lake Neufchatel were used to reconstruct a 1,500-year climate history. The authors say Swiss temperatures fell by 1.5 degrees Celsius during the shift from the Medieval Warming to the Little Ice Age. They also note that mean annual temperatures during the Warming were "on average higher than at present."[67]

The Baltic Sea: A sediment core showed a cold-weather period beginning about 1200, characterized by "a major decrease in the [algae cyst] assemblage and an increase in cold water [algae species]."[68] Thomas Andren of the European Union's Baltic Sea System Study Project, Elinor Andren of Sweden's Upsala University, and Gunnar Sohlenius of the Swedish Royal Institute of Technology also found tropical and subtropical marine plankton species in the Medieval Warming sections of the sediment core that cannot be found in the present Baltic Sea. *The Baltic is thus still too cold to support the warmwater marine species it had in the Medieval Warming.*

In the Mediterranean: Near the equator, temperatures don't change a great deal with the 1,500-year climate cycle, but rainfall regimes do. Two different studies documented the 1,500-year climate cycle in southern Spain's rainfall patterns.[69]

The Eastern Mediterranean: In this area, sediments accumulate rapidly and yield highly accurate seabed cores. Bettina Schilman from the Geological Survey of Israel used such proxies as oxygen-18 and carbon-13 isotopes in phytoplankton, titanium/aluminum ratios, iron/aluminum ratios, magnetic susceptibility, and color index to analyze past climates.[70] She says abrupt climatic events occurred 270 years ago and 800 years ago that "probably correlate" with the Little Ice Age and the Medieval Warming. She also notes corroborating evidence of the Medieval Warming in high Saharan lake levels,[71] and high levels in the Dead Sea[72] and the Sea of Galilee,[73] as well as a precipitation maximum at the Nile headwaters.[74]

The Middle East

Oman: On the Arabian Peninsula, Ulrich Neff of Germany's Heidelberg Academy of Science found oxygen-18 isotopes in a cave stalagmite that yielded a very precise record of the 1,500-year climate cycle in the region's monsoon rainfall. Neff says this monsoon record is at least an order of magnitude better than the seabed sediment cores of the Gerard Bond team that produced the first linkage between the 1,500-year cycle and the sun. Neff's oxygen-18 isotopes also indicate an "excellent" correlation with solar activity (as measured through carbon-14 isotopes in tree rings). Neff's team says the monsoons were considerably stronger during the Climate Optimum (5,000 to 7,000 years ago), which also produced an era of heavy rainfall in the Sahel region of Africa, in Arabia, and in India.[75]

The Oman stalagmite's cycles were also in phase with the temperature fluctuations recorded in Greenland ice cores 10,000 years ago, indicating that both were then controlled by glacial boundaries. For the past 7,000 years, however, since most of the ice melted, the stalagmite proxy says the Indian Ocean monsoon has been governed by variations in solar activity instead.[76]

In the Arabian Sea: West of Karachi, Pakistan, two seabed sediment cores date back nearly 5,000 years and show "the 1,470-year cycle previously reported from the glacial-age Greenland ice record." W. H. Berger and Ulrich von Rad, suggest the cycles were tide driven. However, they also note that *"internal oscillations of the climate system cannot produce them"*[77] (emphasis added).

Asia

In Siberia: Figure 9.2 shows a continuous 2,200-year temperature record from tree rings. We can see the cold period before the Roman Warming, the

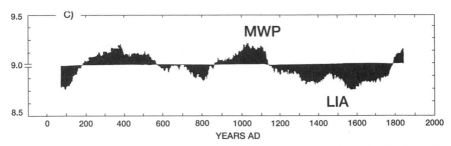

Figure 9.2.   Two Thousand Years of Siberian Winters from Tree Rings Clearly Show the 1,500-Year Cycle. The authors have used a three-hundred-year smoothing of the data.
Source: M. M. Naurzbaev and E. A. Vaganov, "Variation of Early Summer and Annual Temperature in East Taymir and Putoran (Siberia) over the Last Two Millennia Inferred from Tree Rings," *Journal of Geophysical Research* 105 (2000): 7317–26.

Roman Warming itself, the cold Dark Ages, the Medieval Warming extending from 850 to 1150, followed by a sharp cooling from 1200 through 1800.[78]

The Tibetan Plateau: Based on oxygen-18 isotopes from peat bogs, this area had three severely cold intervals during the Dark Ages, a warm period from 1100 to 1300, and then cold periods again during 1370 to 1400, 1550 to 1610, and 1780 to 1880.[79]

In China: a cave stalagmite near Beijing was analyzed using the manganese/strontium ratio as a "geochemical thermometer."[80] The team found a strong warming from 700 to 1000, corresponding to the Medieval Warm Period. From 1500 to 1800, the air temperature was about 1.2 degrees Celsius lower than now.

On Yakushima Island off southern Japan: Carbon-13 isotopes from a giant Japanese cedar infer a warm period about 1 degree Celsius above present levels between 800 and 1200 and 2 degrees Celsius below the present from 1600 to 1700.[81] (The timing of global climate cycles differs between China and Japan due to the fact that ocean currents have a stronger influence on the Japanese islands than they do on the Chinese land mass.)

North America

Let us next examine the two most broadly powerful pieces of evidence of the climate cycles in North America.

First, the North American Pollen Database reveals nine continent-wide temperature-driven shifts in vegetation during the past 14,000 years, an average of one every 1,650 years.[82] The vegetation shifts occurred across the whole of North America! Researchers analyzed more than 3,000 carbon-14 dates from the pollen records, focusing on pollen assemblages just before and after significant climate change periods. The most recent major shift happened about 600 years ago "culminating in the Little Ice Age, with maximum cooling 300 years ago." The previous shift began about 1,600 years ago, and "culminated in the maximum warming of the Medieval Warm Period 1000 years ago."[83]

Second, the water levels of the Great Lakes clearly show a strong response to the 1,500-year climate cycle, with the lake levels high during climate coolings and low during warming periods. Evaporation rates would certainly have been lower during the cool-climate phases, and rainfall might also have been higher. (Ice sheets form only during major ice ages.) Todd Thompson of Indiana University and Steve Baedke of James Madison University studied "strandplains"—shore-parallel sand ridges that have a core of water-laid sediment. These "strandplains" commonly occur on shore as a series of ridges that reveal the upper level of the lake waters and swales, while the organic sediments indicate the age of the ridges. Water levels rose between 3,100 and 2,300 years ago during the cold that preceded the Roman Warming. The lakes were low from 2,300 to 1,900 years ago, reflecting the Roman Warming itself. Then water levels rose again from 100 to 900, in response to the cold Dark Ages.[84] The waters were high again in both 1300 and 1600, reflecting the two-stage Little Ice Age.[85]

In southern Ontario: Ian Campbell and John McAndrews of Environment Canada did a pollen study of successive forest changes as the Little Ice Age cooled the climate, finding that a predominance of warmth-loving beech trees shifted to a predominance of cold-tolerant oaks, then to cold-adapted pine.[86] In a computer simulation, they also concluded that the changes in temperatures and tree species reduced the total plant mass of the Ontario forests by 30 percent in the depths of the Little Ice Age. They say Ontario forests still have not recovered the full productivity they had during the Medieval Warming.

In the Southern Sierra Nevada Mountains: A study of foxtail pine and western juniper in the southern Sierra Nevada Mountains indicates a warmer period than the twentieth century from 1100 to 1375 and a cold period from 1450 to 1850, corresponding to the Little Ice Age.[87]

A chronology of ring widths in the very long-lived bristlecone pine trees on the California-Nevada border extends back to 3431 B.C. From A.D. 800 to the present century, "its hundred-year averages correlate statistically with the temperatures derived for central England."[88]

South America

Temperatures in Latin America vary with the subregion. The tropical regions vary with cloud cover and altitude. Other subregions vary with altitude, wind direction, and sea surface temperatures. At the southern tip, sea surface temperatures are the dominant factor.

In Peru: Alex Chepstow-Lusty found declining amounts of fossilized pollen in a 4,000-year core from a lake bed—indicating declining rainfall for several centuries after 100 as the Roman Warming gave way to the Dark Ages.[89] After 900, increased pollen indicated greater numbers of plants and warmer temperatures, followed by the Little Ice Age and another pollen decline. (The dates of the Peruvian climate oscillations agreed closely with the McDermott stalagmite in the Irish cave.)

In Argentina: Martin Iriondo, of the University of Santa Fe, says historical records—flood reports, sailor's handbooks, and the like—show the central Argentinean area had more precipitation during the Medieval Warming than today. Temperatures may have been as much as 2.5 degrees Celsius warmer due to a southward shift of the Intertropical Convergence Zone.[90]

Also in Argentina, a study of saline lake sediments from a high volcanic plateau found that rainfall and climate changed sharply when the world shifted from warm to cool and back again. The study team concluded, "The Little Ice Age stands as a significant climatic event in the Altiplano and South America."[91]

Africa, South of the Equator

High on Mount Kenya: Researchers from the Weizmann Institute in Israel climbed to Hausberg Tarn, more than 14,000 feet up the mountainside, with their boat and drilling equipment. They retrieved a six-foot core of sediment that had accumulated on the lake bottom between 2250 B.C. and A.D. 750. The team analyzed the ratio of oxygen isotopes in the algae skeletons (called biogenic opal). When the water was cooler, the opal contained less of the heavier oxygen-18 isotopes.

The largest anomaly was a rapid warming—4 degrees Celsius—between 350 B.C. and A.D. 450, reflecting a warmer climate in equatorial East Africa.[92] Was this the Roman Warming? The Weizmann researchers noted warming during the same period in the Swedish part of Lapland and in the northeastern St. Elias Mountains of Alaska and the Canadian Yukon.

In South Africa: The most important evidence of climate cycles includes a cave stalagmite. Peter Tyson, head of the Climatology Research Group at South Africa's University of Witwatersrand, recently led a team analyzing a stalagmite from a cave in eastern South Africa's Makapansgat Valley. The proxies used were oxygen-18 and carbon-14 isotopes and the color density data in its layers. (In warm years, increased organic matter in the soil created a darker layer in the stalagmite; colder years produced lighter layers.)

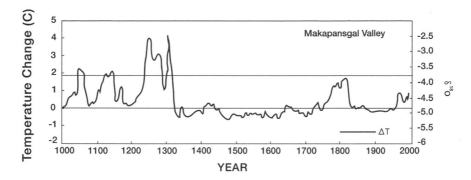

Isotopic Analysis of Stalagmite Samples, South Africa (Tyson et al. 2000)

Figure 9.3. **One Thousand Years of South African Temperatures from a Stalagmite Generally Coincide with Temperature Proxies in the Northern Hemisphere, Showing the Medieval Warming, Little Ice Age, and the Modern Warming.**
Source: Tyson et al., *South African Journal of Science* 96, no. 3 (2000): 121–26.

The Tyson team's temperature record is displayed in figure 9.3. It shows the Medieval Warming starting before 1000 and lasting until around 1300, when temperatures may have been 3 to 4 degrees Celsius higher than at present. The Little Ice Age in the region extended from around 1300 to 1800, when the interior of South Africa was around 1 degree Celsius cooler than today.[93] Temperatures varied during the whole 1,000 years, but varied more during the Medieval Warming. Significantly, the lowest temperatures recorded during the South African Little Ice Age occurred during the Maunder and Sporer Minima when low sunspot counts indicated low solar activity.

The same South African stalagmite revealed cold periods in the cave's region between 800 and 200 B.C.—corresponding to the unnamed cooling period before the Roman Warming.[94] The South African stalagmite doesn't show unusual warming in the twentieth century.

Around the Southern Ocean

In New Zealand: The oxygen-18 isotopes in a stalagmite from a New Zealand cave show exceptionally warm temperatures around 1200 to 1400 followed by the coldest recent period from 1600 to 1700. A. T. Wilson and Chris Hendy of New Zealand's Waikato University state that since their country is "in the Southern Hemisphere and a region meteorologically separated from Europe" finding the Medieval Warming and the Little Ice Age there demonstrates they are "not just a local European phenomenon."[95]

The stalagmite-indicated cold period aligns with the narrowest growth rings in silver pine from the North Island found by Columbia University's D. D'Arrigo.[96]

On Signy Island: Halfway between Antarctica and the southern tip of South America, oxygen isotopes preserved in lake sediments provided a 7,000-year climate record.[97] They clearly show the Roman Warming and the Dark Ages. (Both matched the stalagmite record from McDermott's Irish cave.) Then the lake sediments recorded the Medieval Warming, the Little Ice Age, and the twentieth-century warming—which is cooler to date than the Medieval Warming.

Antarctica

Just off the northern tip of the Antarctic Peninsula, Boo-Keun Khim of South Korea's Seoul University analyzed a sediment core from the Bransfield Basin. The data clearly show the Little Ice Age and Medieval Warming, along with earlier warming/cooling cycles.[98] Khim also notes that evidence of the Little Ice Age was found in several studies of Antarctic marine sediments, including A. Leventer and R. B. Dunbar, who reported on their study of algae microfossils at Antarctica's McMurdo Sound in 1988.[99]

DROUGHTS AND RAINFALL
AS THE CLIMATE WARMED AND COOLED

North America

In Canada: The climate of the western prairies shifted to and from "sustained periods of wetter and drier conditions, occurring approximately every 1220 years" based on the tiny organisms found in the sediments of 219 lakes. The region was dryer during the cold periods, and wetter during the warming periods, with abrupt shifts.[100]

The Western United States: This region suffered at least two century-long droughts during the Medieval Warming that ranged from the Great Plains and Pacific Northwest down into California. The Southeast was also afflicted with droughts. The Upper Midwest and eastern sub-Arctic Canada were much wetter.

In the California Mountains: California geographer Scott Stine analyzed mega-droughts using the radiocarbon dates of volcanic debris layers—supplemented with growth rings from several generations of relict tree stumps and shrubs in the lake shore areas. The trees had been drowned when the lakes regained up to seventy feet of the water depth they had lost during the mega-droughts.[101] Assembling evidence from relict trees in California's Mono, Pyramid, and Tenaya lakes and the Chris Flats of the Walker River, Stine concluded the region had one mega-drought of about 140 years which began around the year 1000, and another century-long drought that began about 1350. Later evidence extended the range of the droughts from Pyramid Lake north of Tahoe to Owens Lake, hundreds of miles to the south, just west of Death Valley. Stine says there were other less-extended, but severe, droughts in the region, including one during the late eighteenth and early nineteenth centuries.[102]

Bristlecone pine tree rings in the California mountains also confirmed a dry, warm period between 700 and 1300.[103]

Stine then points to trees found and dated by geologist Greg Wiles that were sheared off by advancing glaciers in the Prince William Sound region of Alaska—at virtually the same time as the California droughts. Stine suggests that the same "dry-winter-in-California, wet-winter-in-Alaska" model produced by a northward shift of the jet streams in 1976–77 may have prevailed for much of the mega-drought period in the years between 1000 and 1400.

Stine even found trees at least one hundred years old growing in lakebeds in Patagonia, at the southern tip of Argentina, during an extended drought that coincided with at least the first of the California megadroughts. He links the Patagonian drought, as well, to the pattern of jet-stream shift.[104] He seems to be confirming the recently discovered Pacific Decadal Oscillation.

In the Chesapeake Bay: Debra Willard and her team analyzed fossil dinoflagellate cysts and pollen in sediment cores around the Chesapeake. They found several dry periods. One lasted five centuries—from 200 B.C. to A.D. 300—during the Roman Warming. Another dry period lasted from 800 to 1200, during the Medieval Warming.[105] Tree rings[106] and pollen have also shown this period to be drier than average across the southeastern United States.

The Willard team concluded that the extended Mid-Atlantic dry periods generally correspond to "megadroughts" in the central and southwestern United States that "probably exceeded twentieth-century droughts in severity."[107] Their study is corroborated by studies of Great Plains tree rings showing droughts in both the Roman and Medieval Warmings that spanned a century or more;[108] by microfossils and fossilized seed shrimp in lakes;[109] and by dust levels and dust sources in Minnesota lake sediments.[110]

The Willard team found Chesapeake Bay regional droughts in the late sixteenth century that lasted several decades. This was corroborated by another study of tree rings from ancient bald cypress trees in the Virginia tidewater.

The Lost Roanoke Colony in North Carolina: The 117 men, women, and children of the first English settlement in the New World "were last seen on

August 22, 1587, when the tree-ring data indicate the most extreme growing-season drought in 800 years. This drought persisted for 3 years, from 1587 to 1589, and is the driest 3-year episode in the entire 800-year reconstruction. . . . [T]he tree-ring reconstruction also indicates that the settlers of Jamestown Colony had the monumental bad luck to arrive in April, 1607, during the driest 7-year period in 770 years."[111]

Central and South America

Mexico and the Mayans: The famed Mayan civilization in Mexico's Yucatan Peninsula apparently collapsed about 1,200 years ago during an extended drought at the end of the Dark Ages. Sediments from the Cariaco Basin off the coast of northern Venezuela clearly show a drought that started in the seventh century and lasted more than two hundred years. Gerald H. Haug of the University of Southern California and Konrad Hughen of the Woods Hole Oceanographic Institute measured titanium concentrations in the sediments; more titanium was associated with more rainfall.[112]

Mayan cities had thrived in the Yucatán lowlands for 1,000 years—mostly during the Roman Warming era. In the Dark Ages, however, the Mayans suffered at least one hundred years of low rainfall, punctuated by periods of three to nine years in a row with little or no rainfall at all. The driest years were 810, 860, and 910.

Rainfall was critically important to the Maya's because they lived in limestone lowlands where groundwater percolated away rapidly, leaving few lakes and little in the water table to be tapped by wells. The Mayan cities were laid out to catch and store rainfall, and quarries were converted into clay-lined reservoirs. Complex canal systems carried water to irrigate the fields. However, the Maya ultimately depended on rainfall to survive.

The first crisis occurred during the first intense drought of the climate transition period, around 250. Some Mayan cities began to be abandoned. After the rains returned, however, the cities were reoccupied. The Mayan water systems were probably improved and expanded because by 750 the Mayan cities had a total population of more than three million people. That's when the big drought period triggered the final Mayan collapse. The reduced rainfall simply wouldn't support the expanded population.

In Central Chile: Geochemical data, sediments, and algae cyst populations from Laguna Aculeo showed a major increase in floods during 400 B.C. to A.D. 200 (the pre-Roman cold era), 500 to 700 (the Dark Ages), and from 1300 to 1700 (the Little Ice Age). During cooling periods, westerly winds brought additional rainfall to the lake.[113]

The Southern Tip of South America: During the Medieval Warming, and coinciding with a California mega-drought, this area became abnormally dry for several centuries. As we mentioned earlier, Scott Stine found that trees grew as old as one hundred years in the beds of Lakes Argentino, Cardiel, and Ghio before the lakes reflooded during the heavier rains of the Little Ice Age.[114]

Drought in Africa

Africa Is Mostly Tropical or Subtropical: It offers few ancient written records, few mountains and fewer glaciers. It does, however, offer evidence of complex periodic shifts in rainfall patterns.

Working from rainfall records, reports of travelers, local histories and geological studies of lakes and rivers, Sharon E. Nicholson of Florida State University has reconstructed past African climate and environmental changes.[115] She says the "desertification" process that triggered a United Nations conference and much hand-wringing in the early 1980s was "confined to a relatively small scale" and that human impacts on vegetative cover have been far outweighed by natural climate variations. Intense droughts were ubiquitous throughout the period, "long and severe enough to force the migration of peoples and create warfare among various tribes." By the middle of the nineteenth century, with warming, the lakes had returned to very high levels, *"often exceeding 20th-century levels"*[116] (emphasis added).

Nicholson describes a humid period in northern Africa from about the ninth through the fourteenth centuries (the Medieval Warming in Europe) when elephants and giraffe roamed, and towns flourished, in currently dry parts of the Sahara Desert. Major civilizations, such as the Mali Empire, thrived in the Sahel, on the desert's southern border. The Sahel was relatively humid from the sixteenth through the eighteenth centuries.

Farther south, Henry Lamb of the University of Wales led a study of pollen data from the bottom of Kenya's Lake Naivasha.[117] In sharp contrast to the wet climate then prevailing in the Sahara, the team found Kenya had a two-century drought during the warming period, from 980 to 1200. Lamb says the lake's water levels then fell to their lowest point in 1,000 years, while the vegetation shifted strongly away from woody plant species toward grasses.

Sediment cores from Lake Victoria in the Central Highlands show a 1,400 to 1,500 year spacing of precipitation-evaporation fluctuations.[118]

African farming patterns are one of the most obvious proxies for African climate. Thomas N. Huffman of South Africa's University of Witwatersrand used the patterns of millet and sorghum planting as a proxy for climate history in southern Africa.[119] He concluded that carbon-dated cropping evidence essentially proved the southern African climate of the region must have been both warmer and wetter than today from about 900 to 1300. Otherwise the crops could not have been grown where their dated remains have been found.

A recent environmental impact assessment for a gas pipeline linking Mozambique and South Africa reported that the Little Ice Age began about 1300 when its colder and dryer conditions drove the ancestors of the present day Nguni and Sotho-Tswana speakers from East Africa into South Africa. The climate warmed again between 1425 and 1675.[120]

More Global Connections: A remarkable similarity in weather patterns has been documented for southeastern Africa's Lake Malawi and the Cariaco Basin off the coast of Venezuela. Erik T. Brown and Thomas C. Johnson of the University of Minnesota's Large Lakes Observatory studied sediment layers in the north basin of Lake Malawi, and reconstructed a climate record by comparing the algae fossils and ratios of niobium to titanium over 25,000 years.[121]

Knowing that both Lake Malawi and the Caribbean Sea were impacted by the Inter-tropical Convergence Zone, they compared the reconstructed Malawi climate history with the record from the Venezuelan coastal area reconstructed by Gerald Haug based on iron and titanium.

The trends of the two profiles have been "remarkably similar" since the Late Glacial period. Both show the ITCZ being farther north during the Medieval Warming, and farther south during the Little Ice Age. Both records showed more climate variability during the cold period.

SUMMING UP THE WORLDWIDE PHYSICAL EVIDENCE

We have looked at samples of the physical evidence the earth has to offer us, from tree rings and ice cores to stalagmites and dust plumes, from midges and plankton to abandoned prehistoric villages and collapsed cultures, from fossilized pollen and algae skeletons to titanium profiles and niobium ions.

We have assembled evidence from the Arctic and the sub-Arctic, from Europe and China and Tibet, from the equatorial regions of Africa and Latin America, and from the smaller Southern Hemisphere land masses of southern Africa, South America, and New Zealand. We have included important evidence from Antarctica.

None of these pieces of evidence would be convincing in and of themselves. However, in order to dismiss the huge climatic cycle represented by the Medieval Warming and the Little Ice Age, we would have to dismiss not only the human historical evidence from chapter 6, but the enormous range and variety of physical evidence just presented—all of it affirming the 1,500-year climate cycle that goes back at least a million years.

NOTES

1. Lori Stiles, "Medieval Climate Not So Hot," *University of Arizona News*, 16 October 2003.

2. Willie Soon and Sallie Baliunas, "Recent Warming Is Not Historically Unique," *Environment News*, 1 January 2001.

3. Paul Schuster et al., "Chronological Refinement of an Ice Core Record at Upper Fremont Glacier in South Central North America," *Journal of Geophysical Research* 105 (27 February 2000): 4657–66.

4. H. Oeschger et al., "Late Glacial Climate History from Ice Cores," in *Climate Processes and Climate Sensitivity*, ed. J. E. Hansen and Taro Takahashi (Washington, D.C.: American Geophysical Union), Geophysical Monograph #29 (1984), 299–306.

5. C. Lorius et al., "A 150,000-Year Climatic Record from Antarctic Ice," *Nature* 316 (1985): 591–96.

6. The Intertropical Convergence Zone is the region that circles the Earth near the equator. The intense sun and warm water raise the humidity of the air, causing it to rise. Northern and southern movements of the zone drastically affect rainfall in many equatorial nations, producing changes in the wet and dry seasons of the tropics — perhaps with flooding or drought — rather than in the temperature changes of higher latitudes.

7. V. F. Nguetsop et al., "Late Holocene Climatic Changes in West Africa, a High Resolution Diatom Record from Equatorial Cameroon." *Quaternary Science Reviews* 23 (2004)· 591–609.

8. D. Verschuren et al., "Rainfall and Drought in Equatorial East Africa during the Past 1100 Years," *Nature* 403, (2000): 410-414.

9. G. H. Haug et al., "Climate and the Collapse of Maya Civilization," *Science* 299 (2003): 1731–735.

10. S. Huang, H. N. Pollack, and P. Y. Shen, "Late Quaternary Temperature Change Seen in Worldwide Continental Heat Flow Measurements," *Geophysical Research Letters* 24 (1997): 1947–950.

11. J. Esper et al., "1,300 Years of Climate History for Western Central Asia Inferred from Tree Rings," *The Holocene* 12 (2002): 267–77.

12. J. Esper and F. H. Schweingrguber, "Large-Scale Tree Line Changes Recorded in Siberia," *Geophysical Research Letters* 31 (2004): 10.1029/2003GLO019178.

13. L. J. Graumlich, "Global Change in Wilderness Areas: Disentangling Natural and Anthropogenic Changes," U. S. Department of Agriculture Forest Service Proceedings RMRS-P-15-Vol. 3, 2000.

14. F. McDermott et al., "Centennial-Scale Holocene Climate Variability Revealed by a High-Resolution Speleothem O18 Record from SW Ireland," *Science* 294 (2001): 1328–331; S. Niggemann et al., "A Paleoclimate Record of the Last 17,600 Years in Stalagmites from the B7 Cave, Sauerland, Germany," *Quaternary Science Reviews* 22 (2003): 555–67; U. Neff et al., "Strong Coherence between Solar Variability and the Monsoon in Oman between 9 and 6 Kyr Ago," *Nature* 411 (2001): 290–93; and Tyson et al., "The Little Ice Age and Medieval Warming in South Africa," *South African Journal of Science* 96, no. 3 (2000): 121–26.

15. I. D. Campbell and J. H. McAndrews, "Forest Disequilibrium Caused by Rapid Little Ice Age Cooling," *Nature* 366 (1993): 336–38.

16. M. A. Cioccale, "Climatic fluctuations in the Central Region of Argentina in the Last 1000 Years," *Quaternary International* 62 (1999): 35–47.

17. Yang Bao et al., "General Characteristics of Temperature Variation in China during the Last Two Millennia," *Geophysical Research Letters* 10 (2002): 1029/2001GLO014485.

18. B. Wagner and M. Melles, "A Holocene Seabird Record from Raffles So Sediments, East Greenland, in Response to Climatic and Oceanic Changes," *Boreas* 30 (2001): 228–39. The lake (Raffles So) is on the island (Raffles O).

19. R. Monastersky, "Viking Teeth Recount Sad Greenland Tale," *Science News* 19 (1994): 310.

20. W. Soon and S. Baliunas, "Reconstructing Climatic and Environmental Changes of the Past 1,000 Years: A Reappraisal," *Energy & Environment* 14, no. 2/3 (March 2003): 233–96. Willie Soon is a physicist and astronomer at the Mt. Wilson Observatory. Sallie Baliunas is an astrophysicist who has published more than two hundred peer-reviewed scientific papers. She won the Newton Lacey Pierce Prize from the American Astronomical Society and the Bok Prize from Harvard University.

21. M. J. Salinger, "Southwest Pacific Temperatures: Trends in Maximum and Minimum Temperatures," *Atmospheric Research* 37 (1995): 87–100.

22. UN-IPCC, *Third Assessment Report*, 2001 (Cambridge, UK: Cambridge University Press, 2002), chapter 2, section 3.3.

23. E. Cook et al., "Climatic Change in Tasmania Inferred from a 1089-Year Tree-Ring Chronology of Huon Pine," *Science* 253 (1991): 1266–68.

24. Thanks to the late John L. Daly of Tasmania for alerting us to this set of local temperature facts on his long-established climate change Web site, <www.john-daly.com>.

25. John L. Daly, "Talking to the Trees in Tasmania, and Hot Air at Low Head," 1996, <www.john-daly.com> (September 2004).

26. M. J. Salinger, "Southwest Pacific Temperatures: Trends in Maximum and Minimum Temperatures," Atmospheric Research 37 (1995): 87–89.

27. A. T. Wilson et al., "Short-Term Climate Change and New Zealand Temperatures during the Last Millennium," *Nature* 279 (1979): 315–17.

28. Tim Radford, "Glaciers Melting because of Global Warming," *The London Guardian*, 20 February 2001.

29. Alan Cutler, "When Global Cooling Gripped the World," *Washington Post*, 13 August 1997. Cutler was then a visiting scientist at the Smithsonian's National Museum of Natural History.

30. IPCC, *Third Assessment Report*, 128, fig. 2.18.

31. J. Oerlemans et al., "Modeling the Response of Glaciers to Climate Warming," *Climate Dynamics* 14 (1998): 267–74.

32. Dowdeswell et al., "The Mass Balance of Circum-Arctic Glaciers and Recent Climate Change," *Quaternary Research* 48 (1997): 1–14.

33. P. E. Calkin et al., "Holocene Coastal Glaciation of Alaska," *Quaternary Science Review* 20 (2001): 449–461.

34. J. J. Zeeberg and S. L. Forman. "Changes in Glacier Extent on North Novaya Zemlya in the Twentieth Century," *Holocene* 11 (2001): 161–75.

35. Jean. M. Grove, *The Little Ice Age* (Cambridge, UK: Cambridge University Press, 1988).

36. C. J. Caseldine, "The Extent of Some Glaciers in Northern Iceland during the Little Ice Age and the Nature of Recent Deglaciation," *The Geographical Journal* 151 (1985): 215–27.

37. R. J. Braithwaite, "Glacier Mass Balance: The First 50 Years of International Monitoring," *Progress in Physical Geography* 26 (2002): 76–95.

38. R. J. Braithwaite and Y. Zhang, "Relationships between Interannual Variability of Glacier Mass Balance and Climate, "*Journal of Glaciology* 45 (2000): 456–62.

39. C. J. Caseldine, "The Extent of Some Glaciers in Northern Iceland during the Little Ice Age and the Nature of Recent Deglaciation," *The Geographical Journal* 151 (1985): 215–27.

40. A. N. Mackintosh et al., "Holocene Climatic Changes in Iceland: Evidence from Modeling Glacier Length Fluctuation at Solheimajokull," *Quaternary International* 91 (1997): 39–52.

41. T. Yao et al., "Temperature and Methane Records over the Last 2 Ka in Dasuopu Ice Core," *Science in China* (series D) 45 (2002): 1068.

42. Maurizio D'Orefice et al., "Retreat of Mediterranean Glaciers since the Little Ice Age: Case Study of Ghiacciaio del Calderone (Central Apennines, Italy)," *Arctic, Antarctic, and Alpine Research* 32 (May 2000):197–201.

43. D. H. Clark, M. Clark, and A. R. Gillespie, "Little Ice Age Glaciers and Moraines of the Sierra Nevada: Thinly Covered Glacial Ice," in *GSA Abstract with Programs 24* (Oak Ridge, TN: Associated Universities, 1992), 15.

44. G. C. Wiles et al., "Tree Ring Dated Little Ice Age Histories of Maritime Glaciers in Prince William Sound, Alaska," *The Holocene* 9 (1999): 163–73.

45. S. Harrison and V. Winchester, "19th and 20th Century Glacial Fluctuations and Climate Implications in Arco and Colonia Valleys, Hielo Patagonia Norte, Chile," *Arctic, Antarctic, and Alpine Research* 32 (2000): 56–63; A. Y. Goodman et al., "Subdivisions of Glacial Deposits in Southeast Peru Based on Pedogenic Development and Radiometric Ages," *Quaternary Research* 56 (2001): 31.

46. G. Kaser, "A Review of the Modern Fluctuations of Tropical Glaciers," *Global and Planetary Change* 22 (1998): 93–103.

47. G. Kaser et al., "Modern Glacier Retreat on Kilimanjaro as Evidence of Climate Change: Observations and Facts," *International Journal of Climatology* 24 (2004): 329–39.

48. G. Kaser et al., "Modern Glacier Retreat on Kilimanjaro," 329–39.

49. P. Wardle, "Variations of Glaciers of the Westland National Park and Hooker Range," *New Zealand Journal of Botany* 11 (1973): 349–88.

50. S. Winkler, "The Little Ice Age Maximum in the Southern Alps, New Zealand: Preliminary Results from Mueller Glacier," *The Holocene* 10 (2000): 643–647; and J. Purdie and B. Fitzharris, "Processes and Rate of Loss of Tasman Glacier, New Zealand," *Global and Planetary Change* 22 (1999): 79–91.

51. K. Birkenmajer, "Lichenometric Dating of Raised Marine Beaches at Admiralty Bay, King George Island (South Shetland Islands, West Antarctica)," *Bulletin de L'Academie Polonaise des Science* 29 (1981): 119–27.

52. S. Bjorck et al., "Late Holecene Paleoclimate Records from Lake Sediments on James Ross Island, Antarctica," *Palaeogeography, Palaeoclimatology, Palaeoecology* 121 (1996): 195–220.

53. B. Hall and G. Denton, "New Relative Sea Level Curves for the Southern Scott Coast, Antarctica: Evidence for Holocene Deglaciation of the Western Ross Sea." *Journal of Quaternary Science* 14 (1999): 641–50.

54. D. Dahl-Jensen et al., "Past Temperatures Directly from the Greenland Ice Sheet," *Science* 282, (1998): 268–71.

55. A. E. Jennings and N. J. Weiner, "Environmental Change in Eastern Greenland during the Last 1,300 Years: Evidence from Foraminifera and Lithofacies in Nansen Fjord, 68N," *The Holocene* 6 (1996): 171–91.

56. D. Darby et al., "New Record Shows Pronounced Changes in Arctic Ocean Circulation and Climate," *EOS, Transactions, American Geophysical Union* 82 (2001): 601–7.

57. D. R. Muhs et al., "Vegetation and Paleoclimate of the Last Interglacial Period, Central Alaska," *Quaternary Science Review* 20 (2001): 41–61.

58. Z. Gedaloff and D. J. Smith, "Interdecadal Climate Variability and Regime-Scale Shifts in Pacific North America," *Geophysical Research Letters* 28 (2001): 1515–518.

59. J. N. Kasper and M. Allard, "Late Holocene Climatic Changes as Detected by the Growth and Decay of Ice Wedges on the Southern Shore of Hudson Strait, Northern Quebec, Canada," *The Holocene* 11 (2001): 563–77.

60. D. Arseneault and S. Payette, "Reconstruction of Millennial Forest Dynamic from Tree Remains in a Subarctic Tree Line Peatland," *Ecology* 78 (1997): 1873–883.

61. B. E. Berglund, "Human Impact and Climate Changes," *Quaternary International* 105, no. 1 (2003): 7–12.

62. Berglund, "Human Impact and Climate Changes."

63. K. Briffa et al., "A 1,400-Year Tree Ring Record of Summer Temperatures Fennoscandia," *Nature* 346 (1990): 434–39.

64. K. Briffa, "Fennoscandian Summers from A.D. 500: Temperature Changes on Short and Long Timescales," *Climate Dynamics* 7 (1992): 111–19.

65. McDermott, "Centennial-Scale Holocene Climate Variability," 1328–331.

66. Niggemann, "A Paleoclimate Record of the Last 17,600 years," 555–67.

67. M. L. Filippi et al., "Climatic and Anthropogenic Influence on the Stable Isotope Record from Bulk Carbonates and Ostracodes in Lake Neufchatel, Switzerland, During the Last Two Millennia," *Journal of Paleolimnology* 21 (2000):19–34.

68. E. Andren, T. Andren, and G. Sohlenius, "The Holocene History of the Southwester Baltic Sea as Reflected in a Sediment Core from the Bornholm Basin," *Boreas* 29 (2000): 233–50.

69. F. Rodrigo et al., "Rainfall Variability in Southern Spain on Decadal to Centennial Time Scales," *International Journal of Climatology* 20 (2000): 721–32; A. Sousa and P. J. Garcia-Murillo, "Changes in the Wetlands of Andalusia (Donana Nat-

ural Park, SW Spain) at the End of the Little Ice Age, *Climatic Change* 58 (2003): 193–217.

70. B. Schilman et al., "Global Climate Instability Reflected by Eastern Mediterranean Marine Records during the Late Holocene," *Palaeogeography, Palaeoclimatology, Palaeoecology* 176 (2001): 157–76.

71. M. Schoell, "Oxygen Isotope Analysis on Authigenic Carbonates from Lake Van Sediments and Their Possible Bearing on the Climate of the Past 10,000 Years," in *The Geology of Lake Van, Kurtman*, ed. E. T. Degens (Ankara, Turkey: The Mineral and Exploration Institute of Turkey, 1978), 92–97; S. E. Nicholson, "Saharan Climates in Historic Times," in *The Sahara and the Nile*, ed. M. A. J. Williams (Rotterdam, Netherlands: Balkema, 1980), 173–299; Schoell, M., "Oxygen Isotope Analyses on Authigenic Carbonates from Lake Van Sediments and Their Possible Bearing on the Climate of the Past 10,000 Years," in *The Geology of Lake Van, Ankara*, eds. E. T. Degens and F. Kurtman (Ankara, Turkey: Maden Tetkikre Arama Press, 1978), 92–97.

72. A. S. Issar, *Water Shall Flow from the Rock* (Heidelberg, Germany: Springer, 1990); A. S. Issar, "Climate Change and History during the Holocene in the Eastern Mediterranean Region," in *Diachronic Climatic Impacts on Water Resources with Emphasis on the Mediterranean Region*, NATO ASI Series, Series I, Global Environmental Change, ed. A. N. Angelakis and A. S. Issar (Heidelberg, Germany: Springer 1998), 55–75; A. S. Issar et al., "Climatic Changes in Israel during Historical Times and Their Impact on Hydrological, Pedalogical, and Socioeconomic Systems," in *Paleoclimatology and Paleometeorology: Modern and Past Patterns of Global Atmospheric Transport*, ed. M. Leinen and M. Sarnthein (Dordrecht, Netherlands: Kluwer Academic Publishers, 1989), 535–41.

73. A. Frumkin et al., "The Holocene Climatic Record of the Salt Caves of Mount Sedom, Israel," *Holocene* 1 (1991): 191–200; A. S. Issar "Climate Change and History during the Holocene in the Eastern Mediterranean Region," 55–75.

74. B. Bell and D. H. Menzel, "Toward the Observation and Interpretation of Solar Phenomena," *AFCRL F19628-69-C-0077 and AFCRL-TR-74-0357* (Bedford, MA: Air Force Cambridge Research Laboratories, 1972), 8–12; F. Hassan, "Historical Nile Floods and Their Implications for Climatic Change," *Science* 212 (1981): 1142–145.

75. U. Neff et al., "Strong Coherence between Solar Variability and the Monsoon in Oman," *Nature* 411, 290–293 (2001): 7.

76. D. Fleitmann et al., "Holocene Forcing of the Indian Monsoon Recorded in a Stalagmite from Southern Oman," *Science* 300 (2003): 1737–739.

77. W. H. Berger and U. von Rad, "Decadal to Millennial Cyclicity in Varves and Turbidites from the Arabian Sea: Hypothesis of Tidal Origins," *Global and Planetary Change* 34 (2002): 313–25.

78. M. M. Naurzbaev and E. A. Vaganov, "Variation of Early Summer and Annual Temperature in East Taymir and Putoran (Siberia) over the Last Two Millennia Inferred from Tree Rings," *Journal of Geophysical Research* 105 (2000): 7317–326.

79. H. Xu et al., Temperature Variations of the Last 6,000 Years Inferred from O-18 Peat Cellulose from Hongyuan, China," *Chinese Science Bulletin* 47 (2002):1584.

80. Ma Zhibang et al., "Paleotemperature Changes over the Past 3,000 Years in Eastern Beijing, China: A Reconstruction Based on Mg/Sr Records in a Stalagmite," *Chinese Science Bulletin* 48 (2003): 395–400.

81. H. Kitagawa and K. Matsumoto, "Climatic Implications of 13 C Variations in a Japanese Cedar (*Cryptomeria japonica*) during the Last Two Millennia," *Geophysical Research Letters* 22 (1995): 2155–158.

82. The North American Pollen Database is located in Springfield, Illinois, and is part of the Paleoclimatology Program sponsored by the National Oceanographic and Atmospheric Administration.

83. A. E. Viau et al., "Widespread Evidence of 1,500-Yr Climate Variability in North America during the Past 14,000 Years," *Geology* 30, no. 5 (2002): 455–58.

84. T. A. Thompson and S. J. Baedke, "Strandplain Evidence for Reconstructing Later Holocene Lake Levels in the Lake Michigan Basin," in *Proceedings of the Great Lakes Paleo-Levels Workshop: The Last 4,000 Years*, ed. Cynthia Sellinger and Frank Quinn, (Ann Arbor, MI: Great Lakes Environmental Research Laboratory, U.S. Department of Commerce, 1999), 30–34.

85. C. E. Larsen, "A Stratigraphic Study of Beach Features on the Southwestern Shore of Lake Michigan: New Evidence of Holocene Lake Level Fluctuations," *Illinois State Geological Survey Environmental Geology Notes* 112 (1985): 31.

86. Campbell and McAndrews, "Forest Disequilibrium," 336–38.

87. I. D. Graumlich, "A 1,000-Year Record of Temperature and Precipitation in the Sierra Nevada," *Quaternary Research* 39 (1993): 249–55.

88. V. C. LaMarche, "Paleoclimatic Inferences from Long Tree Ring Records," *Science* 183 (1974): 1043–48.

89. A. J. Chepstow-Lusty et al., "Tracing 4,000 Years of Environmental History in the Cuzco Area, Peru, from the Pollen Record," *Mountain Research and Development* 18 (1998): 159–72.

90. M. Iriondo, "Climatic Changes in the South American Plains: Records of a Continent-Scale Oscillation," *Quaternary International* 57–58 (1999): 93–112.

91. B. L. Valero-Garces et al., "Paleohydrology of Andean Saline Lakes from Sedimentological and Isotopic Records, Northwestern Argentina," *Journal of Paleolimnology* 24 (2000): 343–59.

92. M. Rietti-Shati et al., "A 3,000-Year Climatic Record from Biogenic Silica Oxygen Isotopes in an Equatorial High-Altitude Lake," *Science* 281 (1998): 980–82.

93. D. Tyson et al., "The Little Ice Age and Medieval Warming in South Africa," *South African Journal of Science* 96, no. 3 (2000): 121–26.

94. K. Holmgren et al., "A Preliminary 3,000-Year Regional Temperature Reconstruction for South Africa," *South African Journal of Science* 97 (2001): 49–51.

95. A. T. Wilson et al., "Short-Term Climate Change and New Zealand Temperatures during the Last Millennium," *Nature* 279 (1979): 315–17.

96. D. D'Arrigo et al., "Tree Ring Records from New Zealand: Long-Term Context for Recent Warming Trend," *Climate Dynamics* 14 (1998): 191–99.

97. P. E. Noon et al., "Oxygen Isotope Evidence of Holocene Hydrological Changes at Signy Island, Maritime Antarctica," *The Holocene* 13 (2003): 251–63.

98. B. K. Khim et al., "Unstable Climate Oscillations during the Late Holocene in the Eastern Bransfield Basin, Antarctic Peninsula," *Quaternary Research* 58 (2002): 234–45.

99. A. Leventer and R. B. Dunbar, "Recent Diatom Record of McMurdo Sound, Antarctica: Implications for the History of Sea Ice Extent," *Paleooceanography* 3 (1988): 373–86.

100. B. F. Cumming et al., "Persistent Millennial-Scale Shifts in Moisture Regimes in Western Canada during the Past Six Millennia," *Proceedings of the National Academy of Sciences*, USA, 99, no. 25 (2002): 16117–121.

101. S. Stine, "Medieval Climate Anomaly in the Americas," in *Water, Environment, and Society in Times of Climatic Change*, ed. A. S. Issar and N. Brown (Dordrecht, Netherlands: Kluwer Academic Press, 1998), 43–67.

102. S. Stine, "The Great Droughts of Y1K," *Sierra Nature Notes* 1 (May 2001), <www.yosemite.org/naturenotes/paleodrought1.htm>.

103. V. C. LaMarche, "Paleoclimatic Inferences from Long Tree Ring Records," *Science* 183 (1974): 1043–48.

104. S. Stine, "Extreme and Persistent Drought in California and Patagonia during Medieval Time," *Nature* 369 (1994): 546–49.

105. D. A. Willard, "Late-Holocene Climate and Ecosystem History from Chesapeake Bay Sediment Cores, USA," *The Holocene* 13 (2003): 201–14.

106. D. W. Stahle and M. K. Cleaveland, "Tree Ring Reconstructed Rainfall over the Southeastern USA during the Medieval Warm Period and Little Ice Age," *Climatic Change* 26 (1994):199–212.

107. Willard, "Late-Holocene Climate and Ecosystem History from Chesapeake Bay Sediment Cores, USA," 201–14.

108. D. W. Stahle et al., "A 450-Year Drought Reconstruction for Arkansas, United States," *Nature* 316 (1985): 530–32; Stahle and Cleaveland, "Tree Ring Reconstructed Rainfall," 199–212.

109. S. C. Fritz et al., "Hydrologic Variation in the Northern Great Plains during the Last Two Millennia," *Quaternary Research* 53 (2000): 175–84; K. R. Laird et al., "Century-Scale Paleoclimatic Reconstruction from Moon Lake, a Closed-Basin Lake in the Northern Great Plains," *Limnology and Oceanography* 41 (1996): 890–902; and K. R. Laird et al., "Greater Drought Intensity and Frequency before AD 1200 in the Northern Great Plains, USA," *Nature* 384 (1996): 552–54.

110. W. E. Dean, "Rates, Timing, and Cyclicity of Holocene Eolian Activity in North Central United States: Evidence from Lake Sediments," *Geology* 25 (1997): 331–34.

111. Stahle and Cleaveland, "Tree Ring Reconstructed Rainfall."

112. G. H. Haug et al., "Climate and the Collapse of the Maya Civilization," *Science* 299 (2003): 1731–735.

113. B. Jenny et al., "Moisture Changes and Fluctuations of the Westerlies in Mediterranean Central Chile during the Last 2,000 Years: The Laguna Aculeo Record," *Quaternary International* 87 (2002): 3–18.

114. S. Stine, "Extreme Drought in California and Patagonia during Medieval Time," *Nature* 369 (1994): 546–49.

115. S. E. Nicholson and X. Yin, "Rainfall Conditions in Equatorial East Africa during the Nineteenth Century as Inferred from the Record of Lake Victoria," *Climate Change* 48 (2001): 387–98; S. E. Nicholson, "Climatic and Environmental Change in Africa during the Last Two Centuries," *Climate Research* 17 (2001): 123–44.

116. Nicholson and Yin, "Rainfall Conditions in Equatorial East Africa during the Nineteenth Century," 387–98; Nicholson, "Climatic and Environmental Change in Africa during the Last Two Centuries," 123–44.

117. Henry Lamb, I. Darbyshire, and D. Verschueren, "Vegetation Response to Rainfall Variation and Human Impact in Central Kenya during the Past 1,100 Years," *The Holocene* 13 (2003): 285–92.

118. J. C. Stager et al., "A 10,000-Year High Resolution Diatom Record from Pilkington Bay, Lake Victoria, East Africa," *Quaternary Research* 59 (2003): 172–81.

119. T. N. Huffman, "Archeological Evidence for Climatic Change during the Last 2,000 Years in Southern Africa," *Quaternary International* 33 (1996): 55–60.

120. "Mozambique Natural Gas Fields, Environment," 30 December 2003, <www.sasol.com/natural_gas/environment/RSA> (21 October, 2005).

121. E. T. Brown and T. C. Johnson, "The Lake Malawi Climate Record: Links to South America," *Geological Society of America Abstracts with Programs* 35, no. 6 (September 2003): 62.

Chapter Ten

The Baseless Fears

More Frequent and Fiercer Storms

Man-Made Warming Activists Say:

"Europeans have never experienced anything quite like their latest [2000] bout of killer weather . . . unprecedented tandem storms, Martin and Lother, left more than 100 people dead from drowning, avalanches and other weather-related causes. . . . Winds exceeding 200 kilometers per hour broke all existing records in Paris. Baffled meteorologists reported that the force of the winter gales scarcely dissipated as the system swept relentlessly inland from the North Atlantic, spreading a trail of destruction across Britain, France, Germany, Switzerland, Austria and Romania.[1]

"From catastrophic floods to searing droughts, extreme weather linked to global warming has helped send insurance claims soaring in recent years and premiums may follow suit, a study released on Wednesday found. "The extremes around the world are on the increase," Thomas Loster, head of weather and climate risks research at Munich Re, told a news conference with the UN Environmental Programme in Milan. . . . Natural disasters cost the world more than $60 billion in 2003 . . . up from 2002, when global losses were $55 billion."[2]

SEVERE WEATHER IN THEORY AND IN HISTORY

Insurance claims for storm damage are surely one of the poorest ways to measure the impact of global warming. Insurance claims don't measure storm ferocity or frequency. Insurance rebuilds man-made structures. We have more and more expensive structures today and more of them in weather vulnerable locations. Instead, we need to look at the theory and the history of storms.

Earth's weather patterns are driven by the fact that the planet heats unevenly, during the days and through the seasons. The equator, being the broadest part of the planet, absorbs the most heat. The polar regions, having the least surface area, absorb the least. Winds and ocean currents even out the terrestrial temperatures by carrying warm air and warm water away from the equator, and circulating colder air and water back from the polar regions.

The bigger the temperature difference between the equator and the poles, the more power is given to the winds, waves, and currents—and to the storms. Thus, a global warming that primarily increased polar and winter temperatures should mean fewer and milder storms, not bigger, fiercer ones.

In 2003, the UN World Meteorological Organization, which spawned the Intergovernmental Panel on Climate Change (IPCC), issued a statement that global warming could be expected to deliver more "extreme weather events" as carbon dioxide levels rose. However, the IPCC itself had already admitted in its 2001 report that "[n]o systematic changes in the frequency of tornadoes, thunder days or hail events are evident in the limited areas analyzed" and that "changes globally in tropical and extra-tropical storm intensity and frequency are dominated by inter-decadal and multi-decadal variations, with no significant trends evident over the 20th century."[3]

Kerry Emanuel of the respected Massachusetts Institute of Technology has claimed—both in the 1980s and again in 2005—that global warming is producing more powerful and destructive hurricanes. In the summer of 2005, Emanuel charged that global warming had helped to double the destructive power of hurricanes in the North Atlantic and North Pacific during a 0.5 degrees Celsius warming from 1975 to 2005. The new scare was quickly headlined across the country by such media majors as the *Washington Post*, *USA Today*, the *Seattle Times*, and CBS.

Chris Landsea of the NOAA Hurricane Research Division quickly countered that evidence shows natural swings between hurricane highs and lows in periods that often extend for 25 to 40 years. The evidence, which now goes back about 150 years, does *not* show a long-term link with global warming.

One of the nation's leading hurricane forecasters, William Gray of Colorado State University, was harsher with Emanuel's claim. "It's a terrible paper, one of the worst I've ever looked at," said Gray, who denies that cyclone intensity worldwide is increasing. Gray had warned against linking hurricanes and global warming in 2003:

> Some individuals will interpret the recent upswing in hurricane activity during 1995 and 1996 and [my expected higher-than-normal hurricane activity in the next several years] as evidence of climate changes due to increased man-made

greenhouse gases. There is no reasonable way that such interpretations can be accepted.[4]

Moving beyond theory, we turn to history.

One of the most powerful pieces of evidence against Emanuel comes from the Caribbean: historic records tell us that region had nearly three times as many major hurricanes per year during the Little Ice Age from 1701 to 1850 as during the "warming" years from 1950 to 1998. The British Navy was keeping complete and careful records of the storms in the Caribbean from 1700 on, because Britain had big sugar plantations and a large fleet of wooden sailing ships based in the region. Florida State University's James Elsner and his team documented eighty-three major hurricanes hitting Bermuda, Jamaica, and Puerto Rico from 1701 to 1850 (one every two years) and only ten landfalling those islands between 1950 and 1998 (one every five years).[5]

Similarly, a 5,000-year study of tropical hurricanes along a 1,500-kilometer stretch of northeast Australia's coast behind the Great Barrier Reef did not find an increase in "super-cyclones" during the Modern Warming. Jonathon Nott and Matthew Hayne, of the Australian Geological Survey, studied the landform features left by historic hurricanes and then computer-modeled the storms that would have made them.[6] They report that at least five "super-cyclones" with winds over 110 miles per hours hit the Great Barrier Reef in the nineteenth century, but none since 1899.

WHAT ABOUT EXTREME STORM SURGES?

Keqi Zhang of Florida International University led a study, "Twentieth Century Storm Activity along the U.S. East Coast," based on ten very long records of storm surges obtained from tide gauges along the U.S. East Coast. The gauge records were used to calculate indexes of storm duration and intensity. The authors concluded that the record "does not show any discernible long-term secular trends in storm activity during the twentieth century"[7] — despite the global warming in that period.

The Zhang study was supported by the M. E. Hirsch team that studied East Coast winter storms and found no significant long-term trend over forty-six years, beginning in 1951.[8]

Long-term sea level records from several coastal stations in northeastern Europe over the twentieth century show "no sign of a significant increase in storminess" despite the strong warming trend from 1900 to 2000.[9] That's the conclusion of the European WASA Project (Waves and Storms in the North Atlantic).

MAN-MADE WARMING ADVOCATES
LACK KNOWLEDGE OF HISTORY

In testimony before the U.S. Congress, John Christy, director of the Earth System Science Center at the University of Alabama–Huntsville, confirmed that the frequency and severity of droughts, hurricanes, thunderstorms, hail, and tornadoes have *not* increased in recent years. He noted that the most significant droughts in the Southwestern United States occurred more than four hundred years ago, before 1600. He stated that before 1850, America's Great Plains were called "the Great American Desert," and experts at the time said the region couldn't be farmed. Weather just seems unusual and dangerous these days, said Christy, because of the increased media coverage of major storms. The geologic record clearly shows, he says, that today's climate is in no way extraordinary or identifiably different from what could be expected from entirely natural processes.[10]

In contrast, Hubert H. Lamb says more ferocious storms were the "first symptoms" of the rather sudden climatic shift into the Little Ice Age. He notes the "increased incidence and severity of wind storms and sea floods in the thirteenth century. Some of the latter caused appalling loss of life, comparable with the worst disasters in Bangladesh and China in recent times."[11] Remember the three sea floods of the Dutch and German coasts in the thirteenth century mentioned in chapter 7. The death toll was estimated at around 100,000 per flood.

China's Extreme Droughts and Floods

China's written records go back farther and in more detail than those of any other nation—and China has always been particularly subject to droughts and floods. Chinese economist Kang Chao reports that China averaged less than four major floods per century during the Medieval Warming and twice that many during the Little Ice Age. He reports that major droughts were only one-third as common (three per century) during the warm centuries as during the cold centuries (thirteen per century).[12]

NATURAL HAZARDS' SPECIAL ISSUE ON EXTREME WEATHER

The peer-reviewed journal *Natural Hazards* published a special issue on extreme weather events and global warming in June 2003.[13] We'll look at specific authors' research—published in this journal as well as authors published elsewhere—in some detail, but their findings essentially fail to confirm the apocalyptic theories of intensified storminess due to increases in either temperature or CO_2.

The journal's special issue offers an overview of the United States and extreme weather by Robert Balling and Randall Cerveny of the Office of Climatology at Arizona State University. Balling serves as a consultant to the UN's Intergovernmental Panel on Climate Change. He says there is *less* extreme weather as the planet warms. The Balling/Cerveny article further notes that the public is three times more likely to see a news story on severe weather today than thirty years ago—despite the lack of any increase in storminess. The media have responded to the activists' predictions that the world will have more unstable weather as temperatures warm. The public has also been led to believe that climate scientists agree that severe weather will increase as greenhouse gases build up.[14]

Balling and Cerveny say researchers have been "unable to identify significant increases in overall severe storm activity as measured in the magnitude and/or frequency of thunderstorms, hailstorms, tornadoes, hurricanes, and winter storm activity across the United States. There is evidence that heavy rainstorms have increased during the period of historical records. . . . Damage from severe weather has increased over this period, but this upward trend disappears when inflation, population growth, population redistribution, and wealth are taken into account."[15]

Stanley A. Changnon of the University of Illinois and his son David, from Northern Illinois University, checked the number of "thunder days" recorded by the three hundred major U.S. weather stations from 1896 to the present. The number of days with thunder trended upward from 1896 to a peak between 1936 and 1955, and then trended downward moderately after 1955.[16]

The Changnons also examined long-term trends in U.S. hailstorms. Using carefully screened records from sixty-six U.S. weather stations, they found an overall increase in hailstorms from 1916 to a peak around 1955, and then a general decline in hail activity.[17] Hailstorms, in other words, had the same pattern as thunderstorms, and neither agreed with the global climate models' predictions.

U.S. tornado reports have increased by a factor of ten in the last fifty years, but Balling and Cerveny say plotting only the severe tornadoes (F3 and above) shows no upward tornado trend. The increase in tornado reports seems due to better reporting on weaker tornadoes. In fact, Tom P. Grazulis, director of the private sector Tornado Project, warns, "[A]ny link of tornado activity with climatic change of any kind should be treated with the greatest skepticism. The ingredients that go into the creation of a tornado are so varied and complex that they could never be an accurate indicator of climate change."[18]

There *has* been more rain. Many researchers have found that the amount of rainfall from thunderstorms has increased over most of the United States during the past century. This certainly fits with the increased evaporation occurring as the rising twentieth-century temperatures have produced an "invigorated hydrological cycle."[19] By itself, however, the moderate rainfall increase poses little danger.

Natural Hazards on Heavier Rains

E. Kenneth Kunkel of the Illinois State Water Survey finds that there has been a sizable increase in the frequency of heavy U.S. rains since the 1920s—their frequency now matches the level of heavy rains during the late nineteenth century when the country was emerging from the Little Ice Age. Kunkel says the increase in U.S. rainfall does little to resolve the global warming dispute, since either natural warming or anthropogenic forcing of higher temperatures would increase the water vapor taken up by the Earth's atmosphere.[20]

Natural Hazards on Asian Monsoons

The IPCC said in its 1996 report that "[m]ost climate models produce more rainfall over South Asia in a warmer climate with increasing CO_2." In its 2001 report, the IPCC said, "It is likely that the warming associated with increasing greenhouse gas concentrations will cause an increase in Asian summer monsoon variability and changes in monsoon strength."[21]

An Indian research team, led by R. H. Kripalani of the Indian Institute of Tropical Meteorology, has just looked at the variability of India's monsoons as the planet has warmed since 1871.[22] The IPCC expectation has been wrong. The researchers found a series of distinct epochs of three to ten years, with the rainfall alternating between higher or lower than normal in each. *However, there is no evidence that the strength or variability of the Indian monsoons has been affected by the rising temperatures.*

Moreover, during the 1990s—according to the IPCC the warmest decade of the millennium—the *Indian monsoon variability has drastically declined.* Snow depth over the Eurasian continent has been increasing, apparently due to enhanced evaporation and precipitation driven by higher temperatures.

Natural Hazards on Less Extreme Weather in Canada

Madhav Khandekar, a meteorologist with twenty-five years of experience at Environment Canada, responded to the UN World Meteorological Organization's assertion in a peer-reviewed journal of the American Geophysical Union. His study found no increase in extreme weather events (heat waves, floods, winter blizzards, thunderstorms, hail, tornadoes) anywhere in Canada.

In fact, he concluded that "[e]xtreme weather events are definitely on the decline over the last 40 years."[23]

Khandekar noted that the hottest summers of the twentieth century in Canada occurred during the Dust Bowl years of the 1920s and 1930s, not the 1990s. He even noted that "[t]he observed climate change of the last 50 years is beneficial to most regions of Canada in terms of lower heating costs and an enjoyable climate."[24]

Natural Hazards on African Drought

Southern Africa is a particularly drought-prone part of the world because of its geographic location, its steep terrain changes, its contrasted ocean surroundings, and its atmospheric dynamics. Unfortunately, southern Africa's droughts have become more intense and widespread as global temperatures have risen since the late 1960s, according to the French Center for Climatological Research.[25]

Nicolas Fauchereau finds that southern African droughts are associated with warm and widespread increases in tropical sea surface temperatures. However, under those conditions the southern African droughts become less sensitive to temperature itself and more closely linked to the El Niño–Southern Oscillation cycle. In 1970 to 1988, for example, four of the five driest years were associated with peaks of the ENSO cycle.

It is locally devastating that a poverty-stricken region such as southern Africa is suffering more drought as the global climate warms. However, such warmings have occurred before in history—and long before any substantial human emissions of CO_2.

Even if the current warming were man-made, it is entirely possible that investments in southern African infrastructure would do far more to help the region's population than could comparable spending on fossil fuel replacement. Paved roads with ditches and culverts to prevent washouts; storm-sturdy bridges; for agriculture, drought-resistant crop varieties, wells, and new forage systems for livestock; economic growth (factories, ports); and human health (clean water, malaria suppression) would certainly help Africans cope with bad weather. For a few billion dollars, we could improve southern Africa's standards of living whether or not Modern Global Warming adversely affects that area.

PROTECTING OURSELVES FROM HURRICANES

In recent decades, the U.S. National Weather Service has made enormous strides in finding tropical storms with radar while the storms are still infants, tracking them with radar and aircraft, and warning the communities in their potential paths while there is still time to evacuate. This forecasting and warning has had far more to do with human well-being than any change in storm frequency. Moreover, that would have been true even if the frequency and severity of the storms had been increasing rather than declining.

Hurricanes are causing larger and larger financial damages to First World communities, primarily because more and more people have moved to coastal communities, especially attractive beach areas that are, unfortunately, prone to storm damage.

The U.S. has seen a huge boom in Sun Belt economies and population since the invention of air conditioning, especially in Florida and the Gulf Coast, which are prime storm targets. Many more of the buildings in harm's way now are luxury hotels, fancy condominiums, and expensive beach house rentals, rather than the fishing shacks and travel trailers that used to characterize many beach communities. A recent study done for the Federal Emergency Management Agency estimated that the United States had 338,000 buildings within 500 feet of a coastline.[26] It seems inevitable that the low-cost federal flood insurance being offered to the owners of these structures has strongly encouraged such construction.

In the Third World, coastal warning systems are severely limited by poor communications, poor roads, flimsy buildings, and lack of transport. Poor countries also typically lack emergency equipment, trained personnel, and the financial resources to deal with the aftermath of the storms. Consequently, hurricanes still do devastating damage to Third World economies and the rebuilding from one hurricane may not be complete before the next one hits.

The United States

The geologic record indicates that the whole western part of the United States has normally been drier in earlier periods than during the twentieth century. The water wells dug by the Indians in the High Plains 6,000 years ago indicate a drought that lasted 2,500 years, leaving the Great American Desert in its wake.[27] That was during the very warm (Holocene) Climate Optimum when summer temperatures were 2 to 4 degrees Celsius higher and precipitation was lower than today.[28]

Additional evidence of higher temperatures and more dryness comes from annual layers of lake sediment from Elk Lake, Minnesota,[29] and Chappice Lake, Alberta.[30] Both sets of lake sediments indicate a prairie landscape that peaked in warmth and dryness 7,000 years ago during the Climate Optimum. It had only sparse forage and drought-resistant plants and animals.

High Midwest temperatures and lower rainfall also produced extensive sand dunes across much of the western United States, including the High Plains,

Rocky Mountain Basin, the Midwest, Texas, and New Mexico.[31] Explorers on the Great Plains during the nineteenth century wrote of a landscape full of moving sand dunes and sheets.[32] These written records are endorsed by tree ring evidence from the Great Plains in that period.[33] The mobilization of sand during the last 1,000 years has been small, regional, and intermittent by comparison with that of the Climate Optimum.[34] (The prevalence of lightning-caused prairie fires may also have contributed to the bare earth and resulting sand movements.)

Droughts are a potential problem during global warming, but they are also a frequent problem during global cooling.

The flood history of the upper Mississippi Valley includes a warm, dry climate from 3000 B.C. to 1300 B.C., with rain and snowfall totaling about 15 percent less than today, according to James C. Knox of the University of Wisconsin. Then precipitation gradually increased to a period of record floods during the Medieval Warming (900 to 1300) and the transition to the Little Ice Age in the early fourteenth century.[35]

Knox used the flood-deposited cobblestones found in overbank flood plains to calculate the minimum flood depth needed to transport the stones, along with radiocarbon dating to determine the ages of the flood gravels. He says Mississippi floods during the past 150 years have been relatively small and infrequent, with only three extreme floods (1851, 1973, and 1993). If the current global warming repeats the weather patterns of the very warm Climate Optimum (9000 B.C. to 2000 B.C.) the Mississippi Valley could again have much more frequent big floods. But how much of this earlier flooding was due to the melting of the last ice age ice sheets?

The National Climate Data Center also says the southwestern United States had long periods of drought over the last 2,000 years, but with frequent, extreme floods during the Medieval Warm Period.[36] The first two hundred years of the Little Ice Age saw a few large Southwestern floods, but then frequent large floods at the end of that cold era in the nineteenth century. The Center used tree ring data, confirmed by radiocarbon dating of extreme flood deposits.

THE IMPACT OF HEAVIER RAINS

There is little question that we should begin adapting to the probability of heavier rainfall. The increased ocean evaporation that comes with higher temperatures is already increasing the Earth's total precipitation and this trend is likely to continue. This will mean important increases in water runoff. We will need to better protect the world's millions of acres of cropland from soil erosion.

A *Journal of Soil and Water Conservation* special article by Mark Nearing of the USDA's Agricultural Research Service says the models can't predict local changes. However, he concludes, "we can expect about a 2 percent and

1.7 percent change in runoff and erosion, respectively, for each 1 percent change in total precipitation under climate change."[37]

On the other hand, an elegant piece of soil archaeology by Stanley Trimble of UCLA indicates that farmers in the Upper Mississippi are suffering only 6 percent as much soil erosion currently as they did during the Dust Bowl years of the 1930s—thanks to such management strategies as contour farming and the powerful new conservation tillage systems to protect cropped fields.[38]

Conservation tillage, in particular, cuts soil erosion by 65 to 95 percent. It achieves weed control with herbicides rather than plowing, and is now saving topsoil on hundreds of millions of acres across North and South America, Australia, and South Asia. There may be a public interest in using conservation tillage even more broadly as rainfall continues to increase.

THE IMPACTS OF WEATHER EXTREMES
IN THE COMING CENTURIES

We won't have more and bigger storms. We are likely to have more severe droughts and bigger floods in particular regions as temperatures get closer to the peak levels of the Medieval Warming—but the world has had sizable droughts and floods even during global coolings. The big question: Which regions will get the droughts and floods?

We can't answer this question because the computer models are nowhere near precise enough to tell us which regions will be affected and how. As an example, Richard A. Kerr of *Science* looked at the two sets of model results in *Climate Change Impacts on the United States*, published in 2000 by the U.S. Global Change Research Program. "As much as policy-makers would like to know exactly what's in store for Americans, the rudimentary state of regional climate science will not soon allow it," Kerr wrote in an article titled "Dueling Models: Future U.S. Climate Uncertain."[39]

Should California start building new dams and reservoirs to protect itself against a two-hundred-year drought due to global warming? Should we put more research money into improved ways to desalinate seawater? Will that require nuclear power plants? Should California start shifting out of agriculture and conserving the available water for urban uses? Should we transfer California's population to regions that we think will have more plentiful water, such as the soon-to-be-flooded-again Mississippi Valley? Should we start building bigger dams and more expensive flood control levees on the Mississippi and its tributaries?

We can't answer any of these questions with any degree of confidence. What we can say is that humanity's increasing ability to deal with climate problems through knowledge and technology is our best hope for ensuring human well-being and the protection of the environment in the future. There is no realistic hope of "stabilizing" the planet's temperatures at some human-selected level. Moreover, the storminess problems of the coming warm centuries will be mild compared to the storms of the next "little ice age"—due about the year 2280 (+/− 500 years).

NOTES

1. Scott Johnson, "Blown to the Future," *Newsweek International*, 10 January 2000.

2. "Insurers Grapple with Surging Weather Claims," *Reuters*, 12 December 2003.

3. IPCC, *Third Assessment Report, Summary for Policymakers* (Cambridge, UK: Cambridge University Press, 2001), 2.

4. W. Gray et al., "Extended Range Forecast of Atlantic Seasonal Hurricane Activity and U.S. Landfalling Strike Probability for 2004," Colorado State University, December 2003, <http://hurricane.atmos.colostate.edu/forecasts/2004/dec2004/> (26 September 2005).

5. J. B. Elsner et al., "Spatial Variations in Major U.S. Hurricane Activity: Statistics and a Physical Mechanism," *Journal of Climate* 13 (2000): 2293–305.

6. J. Nott and M. Hayne, "High Frequency of Super-Cyclones along the Great Barrier Reef over the Past 5,000 Years," *Nature* 413 (2001): 508–12.

7. K. Zhang et al., "Twentieth-Century Storm Activity along the U.S. East Coast," *Journal of Climate* 13 (2000): 1748–761.

8. M. E. Hirsch et al., "An East Coast Winter Storm Climatology," *Journal of Climate* 14 (2001): 882–99.

9. W. Bijl et al., "Changing Storminess? An Analysis of Long-Term Sea Level Data Sets," *Climate Research* 11 (1999): 161–72.

10. John Christy, Professor of Atmospheric Science, University of Alabama/Huntsville and a lead author for the UN Intergovernmental Panel on Climate Change before House of Representatives Committee, 13 May 2003.

11. H. H. Lamb, *Climate, History, and the Modern World* (New York: Routledge, 1982), 191.

12. Kang Chao, *Man and Land in China: An Economic Analysis* (Palo Alto, CA: Stanford University Press, 1986).

13. *Natural Hazards* 29, no. 2 (June 2003).

14. *Global Warming? The Great Debate: An Interview with Dr. Robert Balling, Jr.*, 11 August 1998, <http://www.evworld.com/archives/interviews/balling.html> (30 September 2005). R. C. Balling Jr. and R. S. Cerveny, "Compilation and Discussion of Trends in Severe Storms in the United States: Popular Perception or Climate Reality?" *Natural Hazards* 29, no. 2 (June 2003): 103–12.

15. *Global Warming? The Great Debate: An Interview with Dr. Robert Balling, Jr.*

16. Stanley Changnon and David Changnon, "Long-Term Fluctuations in Thunderstorm Activity in the United Sates," *Climatic Change* 50 (2001): 489–503.

17. S. A. Changnon and D. Changnon, "Long-Term Fluctuations in Hail Incidences in the United States," *Journal Climate* 13 (2000): 658–64.

18. T. P. Grazulis, *Significant Tornadoes: 1680–1991/A Chronology and Analysis of Events* (St. Johnsbury, VT: Environmental Films, 1993); H. E. Brooks and C. A. Sowell, "Normalized Damage from Major Tornadoes in the United States: 1890–1999," *Weather and Forecasting* 16 (2001): 168–76.

19. S. Changnon, "Thunderstorm Rainfall in the Coterminous United States," *Bulletin of the American Meteorological Society* 82 (2001): 1925–940; and Thomas R. Karl and Richard W. Knight, "Secular Trends of Precipitation Amount, Frequency, and Intensity in the United States," *Bulletin of the American Meteorological Society* 79 (1998): 233–41.

20. E. K. Kunkel, "North American Trends in Extreme Precipitation," *Natural Hazards* 29 (2003): 291–305.

21. IPCC, *Third Assessment Report* (Cambridge, UK: Cambridge University Press, 2001).

22. R. H. Kripalani et al., "Indian Monsoon Variability in a Global Warming Scenario," *Natural Hazards* 29 (2003): 189–206.

23. M. Khandekar, "Comment on [UN World Meteorological Organization] Statement on Extreme Weather Events," *EOS Transactions, American Geophysical Union* 84 (2002): 428.

24. Khandekar, "Comment on [UN World Meteorological Organization] Statement," 428.

25. N. Faucherau et al., "Rainfall Variability and Changes in Southern Africa during the 20th Century in the Global Warming Context," *Natural Hazards* 29 (2003): 139–54.

26. *Evaluation of Erosion Hazards*, H. J. Heinz Center for Science, Economics, and the Environment, FEMA contract EMW-97-CO-0375, 2000, <http://www.heinzctr.org/publications.htm> (30 September 2005).

27. D. J. Meltzer, "The Parching of Prehistoric North America," *New Scientist* 131 (1991): 39–43.

28. Cooperative Holocene Mapping Project Members, "Climate Changes of the Last 18,000 Years: Observations and Model Simulations," *Science* 241 (1988): 1043–52.

29. W. E. Dean et al., "The Variability of Holocene Climate Change: Evidence from Varved Lake Sediments," *Science* 226 (1984): 1191–194.

30. R. E. Vance et al., "7,000-Year Record of Lake-Level Change on the Northern Great Plains: A High-Resolution Proxy of Past Climate," *Geology* 20 (1992): 870–82.

31. S. L. Forman et al., "Large-Scale Stabilized Dunes on the High Plains of Colorado: Understanding the Landscape Response to Holocene Climates with the Aid of Images from Space," *Geology* 20 (1992): 145–48; D. R. Muhs and P. B. Maat, "The Potential Response of Eolian Sands to Greenhouse Warming and Precipitation Reduction on the Great Plains of the U.S.A.," *Journal of Arid Environments* 25 (1993): 351–61; and W. E. Dean et al., "Regional Aridity in North America during the Middle Holocene," *The Holocene* 6 (1996): 145–48.

32. D. R. Muhs and V. T. Holliday, "Evidence of Active Dune Sand on the Great Plains in the 19th Century from Accounts of Early Explorers," *Quaternary Research* 43 (1995): 198–208.

33. D. Meko et al., "The Tree-Ring Record of Severe Sustained Drought," *Water Resources Bulletin* 31 (1995): 789–801; Muhs and Holliday, "Evidence of Active Dune Sand," 198.

34. R. F. Madole, "Stratigraphic Evidence of Desertification in the West-Central Great Plains within the Past 1,000 Years," *Geology* 22 (1994): 483–86.

35. J. C. Knox, "Climatic Influence on Upper Mississippi Valley Floods," in *Flood Geomorphology*, ed. V. R. Baker, R. C. Kochel, and A. C. Patton (New York: Wiley, 1988), 279–300; J. C. Knox, "Large Increases in Flood Magnitude in Response to Modest Changes in Climate," *Nature* 361 (1993): 430–32.

36. National Climate Data Center, "Developing a Tree-Ring Data Bank to Help Answer Questions about Global Change," 1996, <www.ncdc.noaa.gov/paleo/treering .html> 15 February 2005).

37. M. A. Nearing et al., "Expected Climate Change Impacts on Soil Erosion Rates: A Review," *Journal of Soil and Water Conservation* 59 (2004): 43–50.

38. S. Trimble, "Decreased Rate of Alluvial Sediment Storage in the Coon Creek Basin, Wisconsin, 1975–1993," *Science* 285 (1999): 1244–46.

39. R. A. Kerr, "Dueling Models: Future U.S. Climate Uncertain," *Science* 288 (2000): 2113–114.

Chapter Eleven

How Far Can We Trust
the Global Climate Models?

Models are always wrong—but sometimes they're useful anyway.

—Anonymous

Man-Made Warming Activists Say:

"What is missing in most conversations in the U.S. about global warming is a sense of urgency. . . . Under the President's strategy, it's estimated that emissions will actually increase over the next decade. We're speeding toward a wall, and the president is not only refusing to step on the brake, he's accelerating. . . . When you consider that the increased warming is already threatening to decimate the world's coral reefs, and that we're already seeing the melting of the tundra in Alaska, and that alpine ecosystems are already being squeezed off the tops of mountains, it's not too difficult to reach the conclusion that 'too warm' . . . isn't awfully far from where we already are.'"[1]

History-Based Skeptics Say:

"Models that simulate and forecast global climate don't produce the right wobbles, a new study concludes. Despite immense complexity and sophistication, these computer models fail to capture the fluctuations of atmospheric temperatures over months and years. . . . Armin Bunde of the University of Giessen and colleagues . . . compared the results of seven different climate models (GCMs) against measurements of real atmospheric temperatures . . . a universal mathematical relationship known as a 'power law' describes the correlations between temperature fluctuations over . . . timescales from a few months to ten years or more. Now the researchers have found that existing GCMs do not generate this observed scaling law. Some GCMs produce something that looks a little like it on short timescales, but *they mostly generate temperature fluctuations that are essentially random over timescales of more than two years*"[2] (emphasis added).

"The general circulation models presented by the IPCC in 1990 predicted for the regions near the poles in a CO_2 doubling scenario a rise in temperature of more than 12 degrees C. If this were true, in the last 40 years with their steep increase in CO_2 concentration, a warming trend with a temperature rise of several degrees C should have emerged. The opposite is true. A joint investigation by American, Russian, and Canadian scientists shows that the surface temperatures in the Arctic region observed between 1950 and 1990 are going down."[3]

PINNING OUR FUTURE ON
UNVERIFIED COMPUTER MODELS

The previous chapters of this book have presented new, broadly based evidence that a moderate, natural 1,500-year climate cycle has been superimposed over the ice ages and warmings for at least one million years. We stress that the cycle has been *moderate* throughout history and prehistory, as documented in ice cores, seabed sediments, tree rings, and the like. There is nothing in the physical evidence of the planet's history to support scary Hollywood movies or politically correct cries to "end fossil fuel use!" The scary climate stuff in the physical evidence of Earth's history has been due to extraterrestrial shocks, such as asteroid impacts and comet close-calls.

No honest examination of the current planetary climate should begin without those elements of history and prehistory. But none of that is plugged into the computerized Global Climate Circulation models. The models assume a planet with a stable climatic state, even though the million years of climate history in the ice and seabed cores offer no evidence of any such stability in the past. The climate has always been erratically either warming or cooling.

Even the EPA is cautious about the models:

> Virtually all published estimates of how the climate could change in the U.S. are the results of computer models of the atmosphere known as "general circulation models." These complicated models are able to simulate many features of the climate, but they are still not accurate enough to provide reliable forecasts of how the climate may change. . . . Given the unreliability of these models, researchers trying to understand the future impacts of climate change generally analyze scenarios from several different climate models. The hope is that, by using a wide variety of different climate models, one's analysis can include the entire range of scientific uncertainty.[4]

Still, for purposes of debate, let's give the models their clean slate, uninformed by the past climate changes. Let's look at how well they do their job using the rules they choose to employ.

The Global Circulation Models

Global Circulation Models (GCMs) are the mega-stars of today's climate and environmental research. The GCMs are three-dimensional computer models that attempt to pull together—and project into the future—all the major causes of climate change: jet streams in the upper atmosphere; deep ocean currents; solar radiation reflected back to space by ice sheets and glaciers; changes in vegetation; naturally changing greenhouse gas levels; eddies in the ocean that transfer heat laterally; number, type, and altitude of clouds in the skies; variations in radiant energy coming from the sun; plus hundreds of other factors.

The models work from "first principles" such as the laws of thermodynamics and fluid dynamics, the carbon cycle, the water cycle, and so forth. These first principles are set forth in mathematical equations for a vast number of "grid boxes" representing the earth, air and sea around the globe. The computer generates new values for each box as the model steps forward in time by twenty minutes or an hour, again and again through the simulated years. Potential climate change is "pretended" by rerunning the models with varying levels of the greenhouse gases, aerosols, and other factors assumed to be involved in the climate change.

The models are so complex and massive that they can only be run on supercomputers, which means that only wealthy national governments can afford them. Notable GCMs are located at the U.S. National Aeronautics and Space Administration's Goddard Institute for Space Studies (GISS), the U.S. National Center for Atmospheric Research (NCAR), NOAA's Geophysical Fluid Dynamics Laboratory, and Britain's Hadley Centre.

One of the modelers' key problems is that real-world long-term climate changes are sensitive to small changes in surface conditions or solar radiation—so small in the instant that the humans doing the computer inputs may not pick them up. And, if they do, they don't know how to interpret the cumulative future changes they will bring over centuries.

Modern thermometers and ocean buoys have also helped us find a set of shorter climate cycles, measured in years and decades. These include the North Atlantic Oscillation, the Pacific Decadal Oscillation, and the El Niño–Southern Oscillation. The global models can't yet predict even the timing and consequences of the biggest of these, the Pacific Decadal Oscillation.

Another major problem is that the GCMs provide smoothly varying results—but the records we have from ice cores, geology, and paleontology say past climate changes have often been major and abrupt. Among the non-linear feedback factors are such aspects as recognizing that when the last thin layer of an ice sheet melts, the heat-absorbing qualities of the Earth increases suddenly.

During the Younger Dryas, which occurred as the last ice age was ending about 12,000 years ago, a sudden warming must have occurred. Meltwater from the northern ice sheets broke through an ice dam and apparently overwhelmed the Atlantic conveyor currents, plunging the Earth back into another 1,000 years of ultra-cold. The ice sheet reflectance gives us a clue about the suddenness. But what caused the sudden warming?

The GCMs cannot tell us.

Eight thousand years ago, the Sahara Desert was wetter and smaller than today. Its southern border zone, the Sahel, was three hundred miles farther north than it is now. The climate models say the changes in climate wouldn't have changed rainfall enough to account for the desert's reduction—unless we take vegetative feedback into account. Additional rainfall would have encouraged more vegetation, which would have shaded the ground more effectively from the sun. That would have left more moisture in the soil to encourage more vegetation. But what ended that "constructive" vegetative feedback?

The computers don't know.

As an ice age sets in, the Earth becomes colder and dryer, vegetation is sharply reduced, and forests change to grasslands and then to deserts. Each change alters the moisture levels and heat reflectance. What stops the downward spiral of desertification?

The computers provide no answer.

Over the past 500,000 years, the average period of time between ice ages has increased from about 80,000 years to 100,000 years.[5] Why?

No model can tell us.

To date, far more of what we know about the Earth's climate comes from rocks, fossils, and sediment cores than from computer models.

An Expensive Attempt to Forecast the Unknown

The satellite observations of temperatures in the lower atmosphere (up to 30,000 feet), have not shown a *strong* warming trend since the satellites first began to measure globally in 1979. Nor have the high-altitude balloon readings shown a strong warming trend since they began to get near-global coverage in the late 1950s. The satellites' microwave sensors measure the Earth's temperatures more accurately than do surface thermometers that only cover 20 percent of the Earth.

The satellite readings have recently been corrected for a bias in their "diurnal correction" which adjusts for the fact that the satellites do not have onboard propulsion and, thus, their observations gradually drift to later in the day over the several years of each satellite's lifetime. Roy Spencer of the University of Alabama–Huntsville says the correction raised the satellites global temperature trends for 1979–2005 from +0.09 degrees Celsius per decade to +0.12 degrees Celsius per decade. This is still far below the early predictions of greenhouse warming, far below the current surface thermometer trend, and far, far below the current high end of the GCM predictions.

The satellites' moderate warming trend has also been mirrored in the independent readings from instruments carried aloft by high-altitude balloons.

The satellites and balloons are giving us the most accurate temperature measurements we've had in all history, and they provide strong evidence that the Earth's atmosphere has not warmed strongly in a sixty-year period during which greenhouse gas emissions have hugely outstripped any previous human "pollution."

About 80 percent of the carbon dioxide from human activities entered the air after 1940. That means the warming before 1940 must be largely natural, so the human effects cannot reasonably be considered greater than about 0.1 degrees Celsius per decade—the maximum amount of the warming trend seen since the late 1970s.

A reconstruction of weather-balloon temperature readings—at just two meters above the Earth's surface (1979–1996)—shows a temperature increase of only 0.015 degrees Celsius per decade.[6] Yet what we understand of atmospheric physics suggests that sustained warming at the Earth's surface should be amplified with height in the troposphere, not reduced. How does this moderate trend translate into a strong probability of man-made warming totaling 5.8 degrees Celsius by 2100? Especially when each additional CO_2 increment will have less effect on the climate?

North Dakota State Climatologist John Bluemle put the case succinctly:

Atmospheric physics is quite clear that increasing CO_2 concentrations increases temperature. The best way to demonstrate this is to model the temperature of the atmosphere with all CO_2 removed. It is *very* cold. Then, increase CO_2 by small increments and plot the graph of temperature increase. It is very rapid initially and then flattens out. Doubling CO_2 from today's concentration, holding all other parameters constant, has a "negligible" effect.

Regarding the use of computers to model the climate, it should be pointed out that computer models are not information. They are scientists' ideas set to mathematical "music." Real information is what we can actually measure, and what we are measuring does not indicate a significant human contribution to present climate change.[7]

The scary global warming forecasts depend entirely on theoretical computer models—which so far have not even been able to "hindcast" the weather the world has had actually had.

URBAN HEAT ISLANDS, FARMING, AND
OTHER THERMOMETER-CONFOUNDING FACTORS

Comments:

"Heat islands form as cities replace natural land cover with pavement, buildings, and other infrastructure. . . . Displacing trees and vegetation minimizes the natural cooling effects of shading and evaporation of water from soil and leaves."[8]

"[James] Goodridge [former California State Climatologist] groups the temperature trends according to population density. He found that temperature readings from California counties of more than one million inhabitants show an 'increase in temperature commonly attributed to greenhouse warming (as) 3.14 degrees F. per century' while counties with less than 0.1 million people show essentially zero trend. Significantly, the stations selected for a global compilation all show positive temperature trends."[9]

Out in the countryside of Illinois, soil temperatures from a completely rural site showed a gain of 0.4 degrees Celsius from 1882 to 1952. Temperature series from three nearby small towns (less than 2,000 people) showed an increase in the same period of 0.57 degrees Celsius.[10] The difference, 0.17 degrees Celsius, is more than half of the global mean temperature increase of 0.3 degrees Celsius from 1890 to 1950!

Climate researchers have long recognized that the world's official thermometer readings are artificially high because many of them are located at official buildings and airports in urban heat islands. Even a village of 1,000 people can create a heat island, raising its own temperature by 2 to 3 degrees Celsius.[11]

Estimating Both Urban and Rural Land Use Changes

Meteorologists Eugenia Kalnay and Ming Cai say the impact of urbanization and agricultural land use changes on U.S. surface thermometers may have been twice as large as the "urban heat island" factor assumed by the models—and the IPCC. The researchers conclude that U.S. surface warming could have been overestimated by about 40 percent!

Kalnay says past estimates of urbanization and land use changes have too often been based on rising population counts or satellite measurements of nighttime urban lights. Such urban changes can impact thousands of acres of land. However, they do not include changes in reflectance and soil moisture produced by forest clearing, the shifting of land from pasture to crops, and adding irrigation—which have impacted millions of acres of land.

Kalnay was formerly in charge of modeling for the U.S. Weather Service, and helped create today's more accurate three- and five-day "ensemble" weather forecasts. Now at the University of Maryland, she and Cai assembled a non-surface temperature record from the myriad atmospheric observations in the satellite and high-altitude balloon record. Then they used it to reconstruct the last fifty years of U.S. temperatures "without any land use changes."

The U.S. surface thermometer record implies a warming of 0.088 degrees Celsius per decade over the past fifty years. Kalnay and Cai calculated a non-surface warming of only 0.061 degrees Celsius per decade over the fifty years.

The Kalnay-Cai estimate of land use impacts on our temperature record is nearly five times greater than the urbanization assumptions built into the dataset of the U.S. Historical Climate Network.

If the Kalnay-Cai trend is substituted for the trend used in the U.S. National Assessment of the Impacts of Climate Change,[12] the U.S. temperature trend for the twentieth century drops from about 0.45 degrees Celsius to 0.25 degree Celsius. An increase of 0.25 degrees Celsius isn't statistically significant. (Since the thermometer records show the planet's average temperature from 1961–1990 as about 14 degrees Celsius, even the 0.6 degrees of "official" global warming is within the error margin of the thermometer record—0.7 degrees Celsius.)[13]

Remember even more urgently that the planet's temperature never moves in a straight line.

Kalnay and Cai's work identifies a global problem of separating real climate warming from humanity's extensive land use changes and city building. Their conclusion endorses the tree rings, mountain tree lines, and other proxies which tell us the recent Modern Warming has been less extreme than the alarmists believe—and that the Roman Warming and the Medieval Warming were warmer than today.

WHY ARE RICHER COUNTRIES WARMING MORE?

Since the atmosphere is not warming as the Greenhouse Theory says it should, a team from the Netherlands' National Institute for Space Research used local CO_2 emissions as a simple proxy for the amount of local industri-alization.[14] A. T. Jos de Laat and Ahilleas N. Maurellis then divided the world into "industrialized and non-industrialized" regions and calculated the tem-perature trends on that basis. (This doesn't mean that the CO_2 stays in the in-dustrialized areas; it's spread out remarkably evenly at the trace level of 0.037 percent of the air volume.)

The team found that industrial regions with large CO_2 emissions had sig-nificantly stronger warming trends than nonindustrialized regions. They had more warming than the planet as a whole, both at the surface and in the lower atmosphere above the industrialized surfaces. Temperature trends for the re-gions not producing CO_2 were considerably smaller, or even negligible. De Laat and Maurellis speculate that the observed surface temperature changes might be a result of "local surface heating processes and not related to radia-tive greenhouse gas forcing."[15]

What's the difference? De Laat and Maurellis fed their results into two of the climate models used for the IPCC's *Third Assessment Report* (2001). *In-stead of projecting that more CO_2 emissions would produce still-higher tem-peratures, the computer models projected constant or even declining temper-ature trends for the industrialized regions.*

While they were at it, de Laat and Maurellis pointed out a flaw in the IPCC's surface temperature record: It includes virtually no data on the cool-ing of Antarctica in recent decades. They estimate IPCC's selection of re-porting stations—and its mistaken use of a warming trend in the Antarctic— has produced a global warming trend about one-third too high.

Assessing the Surface Temperature Data

Why should we not place full confidence in the surface thermometer records?

The Kalnay-Cai paper argues that global temperatures since 1940 have in-creased only about half as much as the IPCC had assumed. Studies of wealth/industrialization impacts and of rural versus urban temperature records both argue that contamination of the thermometer records has not been fully corrected.

The de Laat and Maurellis computer models seem to say that our official temperatures may be recording a localized surface heating process that will have little to do with the Earth's long-term climate cycle.

We must discount the official temperature history further for the IPCC's failure to include Antarctic cooling data.

The relatively weak warming trends in the satellite and balloon tempera-ture records seem to endorse any or all of these reasons to doubt that the planet will warm radically.

WE REALLY DON'T KNOW HOW TO MODEL CLOUDS

Meanwhile, the latest report of the Intergovernmental Panel on Climate Change itself tells us that the IPCC's consulting scientists don't know how to model the vitally important matter of clouds and their effect on the Earth's temperatures. The report says the climate modelers often don't even know whether a given cloud factor increases or decreases warming, let alone by how much. According to the IPCC's 2001 report:

> In response to any climate [disturbance], the response of cloudiness thereby introduces feedback whose sign and amplitude are largely unknown. . . . Cloud optical feedbacks produced by these GCMs . . . differ both in sign [direction] and strength. The transition between water and ice may be a source of error, but even for a given water phase the sign of the variation of cloud optical properties with temperature can be a matter of controversy.[16]

Roger Davies of NASA points out the wide variety of cloud factors that must be taken into account: shapes, sizes, vertical and horizontal locations, lifetimes, numbers of liquid droplets of different sizes, numbers of ice crystals of different shapes and sizes, and more. He notes that the climatic effect of clouds is so complicated that the leading climate models provide conflicting answers on their impact.[17]

Comments:

"The tropical Pacific Ocean may be able to open a 'vent' in its heat-trapping cirrus cloud cover and release enough energy into space to significantly diminish the projected climate warming caused by a buildup of greenhouse gases in the atmosphere. If confirmed by further research, this newly discovered effect—which is not seen in current climate prediction models—could significantly reduce estimates of future climate warming."[18]

"After examining 22 years of satellite measurements, NASA researchers find that more sunlight entered the tropics and more heat escaped to space in the 1990s than in the 1980s. . . . 'Since clouds were thought to be the weakest link in predicting future climate change from greenhouse gases, these new results are unsettling,' said Dr. Bruce Weilicki of NASA. . . . 'It suggests that current climate models, may, in fact, be more uncertain than we had thought.' The previously unknown changes in the [Earth's] radiation budget are two to four times larger than scientists had believed possible . . . several of the world's top climate modeling research groups agreed to take on the challenge of reproducing the tropical cloud changes. But the climate models failed the test, predicting smaller than observed variability by factors of two to four, . . . said Weilicki, 'We tracked the changes to a decrease in tropical cloudiness that allowed more sunlight to reach the Earth's surface. But what we want to know is why the clouds would change.'"[19]

After the NASA studies revealed the massive planetary heat vent over the Pacific, the agency threw the "heat vent" anomaly into the laps of the global climate modeling community. NASA asked the climate modelers to reconfigure their models to account for this previously unknown and newly observed reality. They were unable to do so.

As noted earlier in this book, this was a tremendously important event (or should have been) in the global climate debate. After all, the Pacific's vertical heat vent emitted about as much energy during the 1980s and 1990s as would have been generated by an instant doubling of the air's CO_2 content.

NASA's Bruce A. Weilicki wrote, "We present new evidence from a compilation of over two decades of accurate satellite data that the top-of-atmosphere tropical radiative energy budget is much more dynamic and variable than previously thought. Results indicate that the radiation budget changes are caused by changes in tropical mean cloudiness. The results of several current climate model simulations fail to predict this large observed variation in tropical energy budget. The missing variability in the models highlights the critical need to improve cloud modeling. . . . This leads to a threefold uncertainty in the predictions of the possible global warming over the next century. . . . We caution against interpreting the decadal variability as evidence of greenhouse gas warming."[20]

The mainstream media essentially ignored the heat vent discovery, along with Weilicki's caution that this discovery calls into question the whole greenhouse explanation.

We Can't Trust the Climate Models at All

The Earth's surface thermometers are heavily skewed by urban heat and land use changes; they may overstate U.S. surface warming by a factor of 40 percent.

The models have erroneously predicted a twentieth-century surge in Earth's temperatures to match surging CO_2 concentrations in the atmosphere. It hasn't happened. The degree of climate forcing assumed in the still-unverified models is apparently far too high.

The Greenhouse Theory also seems to be failing at the poles, where the warming was supposed to be earliest and strongest. The models say we should have seen warming of several degrees Celsius at the poles since 1940, to reflect the major increases in atmospheric CO_2. Instead, polar temperatures *have been falling*.

However, the biggest failing of the Greenhouse Theory may be in the troposphere, which seemingly should warm faster than the Earth's surface. The troposphere has warmed much less than the Earth's surface. Nor do the models have any explanation for the massive tropical heat vent over the Pacific, which is seemingly capable of negating a doubling of CO_2.

What explanation does the IPCC offer for these major departures from its basic man-made warming contention?

None.

The American Association of State Climatologists has issued a policy statement on climate change that says, essentially, we can't predict the planet's temperature change:

POLICY STATEMENT ON CLIMATE VARIABILITY AND CHANGE

1. Past climate is a useful guide to the future. . . .
2. Climate prediction is complex with many uncertainties—the AASC recognizes climate prediction is an extremely difficult undertaking. For time scales of a decade or more, understanding the empirical accuracy of such predictions—called "verification" is simply impossible, since we have to wait a decade or longer to assess the accuracy of the forecasts.
3. Policy responses to climate variability and change should be flexible and sensible . . . the AASC recommends that policies related to long-term climate not be based on particular predictions. . . .
4. [T]he nation's climate policies must involve much more than discussions of alternative energy policies. . . . The climate data must include all important components of the climate system (e.g., temperature, precipitation, humidity, vegetation health, and soil moisture)."[21]

NOTES

1. Bob Herbert, "How Hot Is Too Hot?" *New York Times*, 24 June 2002.

2. Philip Ball, "Shake-up for Climate Models," *Nature*, 1 July 2002, <www.nature.com/nsu/020624/020624-11.html>.

3. Theodor Landscheidt, *Solar Activity: A Dominant Factor in Climate Dynamics* (Nova Scotia: Schroeter Institute for Research in Cycles of Solar Activity, 1998).

4. U.S. Environmental Protection Agency, *Global Warming—Climate*, 14 October 2004, <www.yosemite.epa.gov/introduction/goals>.

5. J. A. Rial et al., "Nonlinearities, Feedbacks, and Critical Thresholds within the Earth's Climate System," *Climate Change* 65 (2004): 11–38.

6. D. H. Douglass, B. D. Pearson, and S. F. Singer, "Altitude Dependence of Atmospheric Temperature Trends: Climate Models versus Observations," *Geophysical Research Letters* 31 (2004): L13208, doi: 10 (1029/2004): GL020103; and D. H. Douglass et al., "Disparity of Tropospheric and Surface Temperature Trends: New Evidence," *Geophysical Research Letters* 31 (2004): L13207, doi: 10 (10292004): GL020212.

7. John Bluemle, "Some Thoughts on Climate Change," *North Dakota State Geological Survey Newsletter* 28 (2001): 1–2.

8. U.S. Environmental Protection Agency, "Heat Island Effect: Basic Information," <http://www.epa.gov/heatisland/about/index.html>, 7 February 2006.

9. S. Fred Singer, *Hot Talk, Cold Science* (Oakland, CA: The Independent Institute, 1997), 47. Singer quotes James D. Goodridge, "Urban Bias Influences on Long-Term California Air Temperature Trends," *Atmospheric Environment* 26B 1 (1992): 1–7.

10. S. A. Changnon, "A Rare Long Record of Deep Soil Temperatures Defines Temporal Temperature Changes and an Urban Heat Island," *Climatic Change* 38 (1999): 113–28.

11. T. R. Oke, "City Size and the Urban Heat Island," *Atmospheric Environment* 7 (1987): 769–79.

12. *U.S. National Assessment of the Potential Impacts of Climate Variability and Change*, U.S. Global Change Research Program, Washington, D.C., 2000

13. E. Kalnay and M. Cai, "Estimating the Impact of Urbanization and Land Use on U.S. Surface Temperature Trends: Preliminary Report," *Nature* 423 (29 May 2003): 528–31.

14. A. T. J. de Laat and A. N. Maurellis, "Industrial CO_2 Emissions as a Proxy for Anthropogenic Influence on Lower Tropospheric Temperature Trends," *Geophysical Research Letters* 31 (2004): L05204, doi: 10 (1029/2003): GL019024.

15. de Laat and Maurellis, "Industrial CO_2 Emissions as a Proxy."

16. *Climate Change 2001*, chapter 7, section 7.2.2.4: "Cloud Radiative Feedback Processes."

17. Roger Davies, *Science Goals: Study of Clouds*, NASA, May 2004, <www-misr/jpl.nasa.gov/mission/introduction/goals3.html>.

18. "Natural 'Heat Vent' in Pacific Cloud Cover Could Diminish Greenhouse Warming," American Meteorological Society news release, 28 February 2001.

19. "Fewer Clouds Indicate Climate Change," 1 February 2002, <www.Science aGoGo.com>.

20. B. A. Weilicki et al., "Evidence for Large Decadal Variability in the Tropical Mean Radiative Energy Budget," *Science* 295 (2002): 841–44.

21. American Association of State Climatologists, November 2001, <lwf.ncbc .noaa.gov/oa/aasc/AASC_on_Climate.htm>.

Chapter Twelve

The Baseless Fears

Abrupt Global Cooling

Comments:

"Today's global warming could very well destabilize the North Atlantic current, triggering a return to Ice Age climate conditions in the Northern Hemisphere and much of the rest of the world. Agriculture worldwide would be profoundly impacted. Distribution of what food remained would be hugely disrupted. People in cities would starve. After a catastrophic downsizing of the population we'd be left with a lot of little despotic governments . . . a world full of Yugoslavias."[1]

"When I say 'dramatic'' I mean: Average winter temperatures could drop by 5 degrees Fahrenheit over much of the United States. . . . That's enough to send mountain glaciers advancing down from the Alps. To freeze rivers and harbors and bind North Atlantic shipping lanes in ice. To disrupt the operation of ground and air transportation. To cause energy needs to soar exponentially. To force wholesale changes in agricultural practices and fisheries. . . . It is reasonable to assume that greenhouse warming can exacerbate the possibility of precipitating large, abrupt and regional or global climatic changes. We even have strong evidence that we may be approaching a dangerous threshold — that we are squeezing a trigger in the North Atlantic."[2]

"Approximately 12,800 years ago, as the climate was warming following the Earth's last glacial maximum ("ice age") an abrupt transition to cold conditions occurred, during which the surface temperature of the Northern Hemisphere dropped precipitously (nearly 27 degrees F. in Greenland, for example) in a series of abrupt, decadal-scale jumps. . . . This abrupt climate cooling is known as the "Younger Dryas," event . . . the Northern Hemisphere, especially Europe and Greenland, experienced considerably colder conditions lasting about 1,300 years. . . . The termination of this cold event around 11,500 years ago occurred as an even more abrupt warming, most of which took place in a single, 5-year period."[3]

CAN WE THINK OF FOUR REASONS
NOT TO TAKE THIS TOO SERIOUSLY?

First, consider the trillions of tons of glacial ice we don't have because it melted more than 5,000 years ago.

Twelve thousand years ago, the Gulf Stream did get overwhelmed—by melt-water from the huge ice sheets and glaciers of the ice age as the planet warmed into the Climate Optimum. But the ice age had created an ice sheet up to 9,000 feet thick over the northern part of Europe and North America. The Laurentian Ice Sheet in the center of North America extended over all the Great Lakes, west into Iowa and south into Indiana and Ohio.

We calculate there were thus some 40,000 trillion tons of ice in the world's various ice sheets and glaciers at that time.

Has Chicago recently reported an ice sheet a mile thick covering City Hall?

Second, the recent warming years, far from triggering a shutdown of the Atlantic Conveyor, have produced a rapid and systematic *increase* in the flow rate of the deep Atlantic currents.[4]

Third, Dr. Gagosian, director of the Woods Hole Oceanographic Institute, is clearly engaged in the now-widespread scientific practice of "scaring up research funds." He noted in his statements that the seas are responsible for about half of the global climate factors, while virtually all of the global warming research money has gone to the atmospheric scientists.

Richard B. Alley, chair of the National Research Council's Committee on Abrupt Climate Change, wrote in a report titled *Abrupt Climate Change: Inevitable Surprises* that we face a heightened potential for large and rapid temperature transitions as a result of man-made global warming. The Greening Earth Society recently quoted one of the committee members as saying, "The intent of the report was to draw attention to a common field of interest so that they could raise awareness enough to generate some funding support."[5]

Yet history and the physical evidence both tell us that the warming phases of the climate cycle have been far more stable and pleasant than the cold phases.

Fourth, computerized global circulation models say it won't happen!

Computer modeling can be useful, when the factors involved are known and the instructions to the computer can be based on real-world data.

After the publication of *Abrupt Climate Change, Inevitable Surprises,* researchers at the Lamont-Doherty Earth Observatory ran several versions of the Gulf Stream Collapse theory on the global climate model at the Goddard Institute for Space Studies. The team found no evidence for a "tipping point" that would push the planet from the projected warming of the twenty-first century to an abrupt global cooling. Instead, they found a linear response to glacial meltwater. They say the expected increase in global melt water with the Modern Warming "is not rapid with realistic freshwater inputs."[6] In other words, without an extra 40,000 trillion tons of ice to melt, warming won't shut down the Atlantic Conveyor—as it didn't during the warming surges of 1850–1870 or 1920–1940.

Abrupt Climate Change: Inevitable Surprises warns about "large abrupt climate changes" of "as much as 10 degrees C in 10 years," stating that such changes "are not only possible but likely in the future."[7]

The Lamont-Doherty team found no evidence for this sort of dramatic "thresholds." David Rind says that in his team's model runs, the Atlantic Conveyor "decreases linearly with the volume of fresh water added through the St. Lawrence" and it does so "without any obvious threshold effects."[8]

Nor did they find any evidence for a huge increase in fresh water inputs from the projected warming, despite its related increase in rainfall. One reason, of course, is that snow can accumulate on ice sheets up to four times faster during warm periods than during relatively cold periods.[9] Another reason is that ice melts slowly, deflecting much of the sun's heat away from its core. As a result, say the Rind researchers, the effect of warming on the North Atlantic's deep water convection system "is not rapid."[10]

Peili Wu and a team at the U.K.'s Hadley Centre for Climate Prediction and Research used the Hadley climate model to test the same hypothesis—that increased melt-water could shut down the ocean's circulation.[11] It didn't.

The Hadley model, in fact, found just the opposite: "accompanying the freshening trend, the [thermohaline circulation] unexpectedly shows an upward trend, rather than a downward trend." The model thus agrees with the real-world trend. The deep ocean currents are becoming more vigorous with the warming and its increased precipitation.

NOTES

1. William Calvin, neurophysiologist and author of *A Brain for All Seasons: Human Evolution and Abrupt Climate Change* (Chicago: University of Chicago Press, 2002).

2. Robert Gagosian, "Abrupt Climate Changes. Should We Be Worried?" Paper presented at the World Economic Forum, Davos, Switzerland, 27 January 2003.

3. U.S. Global Change Seminar: "Abrupt Climate Changes Revisited: How Serious and How likely?" U.S. Global Change Research Program Seminar Series, 23 February 1998.

4. R. R. Dickson et al., "Rapid Freshening of the Deep North Atlantic Ocean over the Past Four Decades," *Nature* 416 (2002): 832–37.

5. Richard B. Alley, preface in The Committee on Abrupt Climate Change of the National Academy of Sciences, *Abrupt Climate Change: Inevitable Surprises* (Washington, D.C.: National Academy Press, 2001), v–vi.

6. David Rind et al., "Effects of Glacial Melt Water in the GISS Coupled Atmosphere-Ocean Model 1, North Atlantic Deep Water Response," *Journal of Geophysical Research* 106 (2001): 27335–353.

7. Alley, v–vi.

8. Rind et al., "Effects of Glacial Melt Water in the GISS Coupled Atmosphere-Ocean Model 1, North Atlantic Deep Water Response," 27335–353.

9. E. Tziperman and H. Gildor, "The Stabilization of the Thermohaline Circulation by the Temperature-Precipitation Feedback," *Journal of Physical Oceanography* 32 (2002): 2707–12.

10. Rind et al., "Effects of Glacial Melt Water in the GISS Coupled Atmosphere-Ocean Model 1, North Atlantic Deep Water Response," 27335–353.

11. P. Wu et al., "Does the Recent Freshening Trend in the North Atlantic Indicate a Weakening Thermohaline Circulation?" *Geophysical Research Letters* 31 (2004): 10.1029/2003GL018584.

Chapter Thirteen

The Sun–Climate Connection

Comments:

"For more than a hundred years there have been reports of an apparent connection between solar activity and Earth's climate. William Herschel, a famous scientist in London, suggested in 1801 that the price of wheat was directly controlled by the number of sunspots, based on his observation that less rain fell when there were few sunspots. . . . Solar activity is now known very far back in time due to the production of isotopes in the atmosphere by galactic cosmic rays. From such records there is a striking qualitative agreement between cold and warm climatic periods and low and high solar activity during the last 10,000 years."[1]

"German scientists have found a significant piece of evidence linking cosmic rays to climate change. They have detected charged particle clusters in the lower atmosphere that were probably caused by the [cosmic rays]. They say the clusters can lead to the condensed nuclei which form into dense clouds. . . . The amount of cosmic rays reaching Earth is largely controlled by the Sun, and many solar scientists believe the star's indirect influence on Earth's global climate has been underestimated."[2]

Fifty years ago, scientists spoke of the "solar constant." The sun was the huge, unvarying source of Earth's energy. However, in recent years, isotope variations in ice cores, tree rings, and seabed sediments have confirmed a connection between a 1,500-year cycle in the Earth's climate and a tiny variation in the sun's activity. The variability of the "eleven-year sunspot cycle" is a key part of the evidence.

Figure 13.1 shows the relationship between the length of the solar cycles and Earth's temperatures, as measured in the Northern Hemisphere over the past three hundred years. (The Southern Hemisphere's temperatures are assumed to be closely related but we have no thermometer records.) Given the variability of the temperatures, the close relationship between the two is startling.

Richard Willson, affiliated with both Columbia University and NASA, reported that the sun's radiation has increased by nearly 0.05 percent per decade since the late 1970s. Willson used data from three different NASA ACRIM satellites monitoring the sun to assemble a twenty-five-year record of total solar irradiation from 1978 to 2003. The trend is significant because the sun's total energy output is so huge. A variation of 0.05 percent in its output is equal to all human energy use. Willson says he can't be sure that the trend toward increasing solar radiation goes back further than 1978, but says that if this trend had persisted throughout the twentieth century, it would have produced "a significant component" of the observed warming.[3]

Here is the sun–climate hypothesis—that tiny variations in the sun's irradiance are amplified into major climate changes on Earth by at least two factors: (1) cosmic rays creating more or fewer of the low, cooling clouds in the Earth's atmosphere; and (2) solar-driven ozone changes in the stratosphere creating more or less heating of the lower atmosphere.

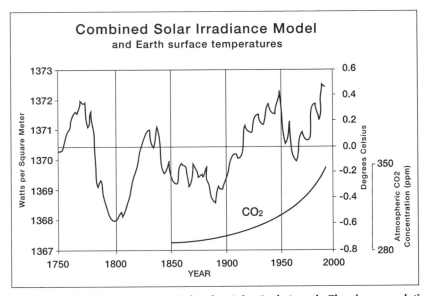

Figure 13.1.   Earth's Temperatures Related to Solar Cycle Length. The close correlation endorses the strong relationship between solar irradiance and Earth's climate.

Source: S. Baliunas and W. Soon, "Solar Variability and Climate Change," *Astrophysical Journal* 450 (1995): 896.

CHANGES IN CLOUDS CHANGE OUR CLIMATE

Comment:

"So if the sun undergoes changes in activity—which it does—the amount of cosmic rays reaching the earth will also vary over the same timescale, and so will the planet's overall cloudiness. . . . *Data collected from satellites show that the amount of low clouds over the earth closely follows the amount of cosmic rays reaching the earth*"[4] (emphasis added).

It was not until the 1960s that our satellites could accurately gauge the sun's variation through ultra violet rays out in the clarity of space.

In 1961, the University of Washington's Minze Stuiver (called the second-most-cited scientist in geosciences) was able to publish a correlation between solar activity and carbon-14 variations in the tree-ring records of the past 1,000 years.[5] He concluded that when an active sun created more solar wind to shield the Earth from cosmic rays less carbon-14 was available to be absorbed by the trees.

Charles Perry of the U.S. Geological Survey and Kenneth Hsu recently took this sun–tree-ring connection through an entire 90,000-year glacial cycle, with a solar-luminosity model based on the variations in the solar-related isotopes. They found the model correlated well with the carbon-14 readings from tree rings, carefully dated back as far as the Medieval Warming.

Perry and Hsu finally conclude that the idea of "the modern temperature increase being caused solely by an increase in CO_2 concentrations appears questionable."[6]

Drew Shindell of NASA's Goddard Institute for Space Studies used Goddard's global climate model to compare the earth's climate during the Maunder Solar Minimum (1645 to 1710) with a period one hundred years later when solar output was relatively high for several decades. Shindell found the increased solar activity led to only a modest warming (about 0.35 degrees Celsius) for the Earth as a whole over the one hundred years after the Maunder Minimum.

However, the modest overall warming was amplified into a five times greater temperature response for the Northern Hemisphere continents in winter; temperatures in northern Europe increased by 1 to 2 degrees Celsius. The researchers say their model results tally closely with the physical evidence from both the Maunder Solar Minimum and the period of a more active sun a century later—as preserved in tree rings, ice cores, corals, and historical records.[7]

The Cosmic Ray/Cloud Cover Amplifier

Comment:

"Svensmark in 1997 . . . showed that there is a correlation between total cloud cover over the Earth and the influx of cosmic rays. . . .[8] The rays are thought to collide with particles or molecules in the atmosphere, leaving them electrically charged, or 'ionized.' These ionized particles then seed the growth of cloud water droplets. . . . Clouds that form low in the sky are relatively warm and made up of tiny water droplets. These tend to cool the planet by reflecting sunlight back into space. High clouds are cooler, consisting mostly of ice particles, and they can have the opposite effect. . . . By studying satellite measurements . . . since 1980, Svensmark and Marsh have found that only low-altitude clouds . . . seem to vary in step with the rise and fall of the cosmic-ray flux. They argue that the imprint of the solar magnetic field in the solar wind has increased over the past century. So the shielding from cosmic rays will have increased, decreasing the formation and cooling influence of low clouds, and providing a possible contribution to the observed global warming."[9]

The sun constantly releases a stream of charged particles—the solar wind—which partially shields the earth from cosmic rays. This solar wind varies with the sun's irradiance.

When the sun's activity is weak and the solar wind blows less forcefully, more cosmic rays streak through our atmosphere, creating more low clouds that, in turn, increase the Earth's ability to reflect more of the sun's visible-range heat away from the planet. That's a cooling effect. (That's why cloudy skies predominate in landscape paintings from the Little Ice Age.)

When the sun is stronger, as it has been recently, the solar wind blows more strongly, and the Earth is shielded more effectively from the cosmic rays. That means fewer low clouds and more warming.

Svensmark matched the data on cosmic rays from the neutron monitor in Climax, Colorado, with the satellite measurements of solar irradiance from 1970 to 1990. Over the period between 1975 and 1989, he found cosmic rays increased by 1.2 percent annually, amplifying the sun's change in irradiance about fourfold. He concluded that, "The direct influence of changes in solar irradiance is estimated to be only 0.1 degree C. The cloud forcing, however, gives for the above sensitivity, 0.3–0.5 C., and *has therefore the potential of explaining nearly all of the temperature changes in the period studied*" (emphasis added). He admitted, "There is at present no detailed understanding of the microphysical mechanism that connects solar activity and earth's cloud cover."[10]

On the other hand, we try to avoid getting hit by a bus even if we can't explain how its diesel engine works.

The Ozone Amplifier

Joanna Haigh of London's Imperial College, presented a paper at the 1998 meeting of the American Association for the Advancement of Science on "The Effects of Change in Solar Ultraviolet Emission on Climate." Haigh said more "far UV" from the sun produces more ozone in the atmosphere—and that ozone then absorbs more of the near-UV radiation from the sun. Her computer modeling study suggested that a 0.1 percent variation in the sun's radiation could cause a 2 percent change in ozone concentration in the Earth's atmosphere.[11]

Ozone is produced mostly in the upper stratosphere by the high-energy ("far") ultraviolet (UV) rays of the sun striking—and splitting—oxygen molecules (O_2). Some of these molecules reform into ozone (O_3), which forms a thin, protective stratospheric shield against most ("near") UV rays striking the Earth itself. The energy absorbed by ozone from the near-UV of the sun heats the stratosphere.

NASA's Shindell says his team confirmed that ozone is one of the key factors that amplify the effects of changes in the sun's irradiance. They ran two identical experiments simulating climate, except they left out the ozone response in one set. They found "an analogous but weaker response without interactive ozone."[12]

It's More than the Sun; It's the Whole Galaxy

Jan Veizer, a geologist at the University of Ottawa, reconstructed the Earth's temperature record for the last 500 million years, using the calcium and magnesium isotopes in fossilized seashells. (Half a billion years is how long the Earth's marine creatures have made seashells.) He was surprised to find a major global warming-cooling cycle every 135 million years, a time period that coincides with no earthly phenomenon.

Then, Nir Shaviv, an astrophysicist at the Hebrew University of Jerusalem, visited Toronto and told Veizer that cosmic rays striking the Earth cycle up and down over 135 million years as our solar system pass through one of the bright arms of the Milky Way. The Milky Way arms have intense levels of cosmic rays that tend to cool the Earth, by stimulating the formation of low-level clouds that reflect solar radiance back into space.

In a cross-disciplinary study recently published by the Geological Society of America, Veizer and Shaviv conclude that 75 percent of the Earth's temperature variability in the past 500 million years is due to changes in the varying bombardment by cosmic rays as we pass in and out of the spiral arms of the Milky Way.[13]

"Our approach, based on entirely independent studies from astrophysics and geosciences, yields a surprisingly consistent picture of climate evolution on geological time scales. The global climate possesses a stabilizing negative feedback. A likely candidate for such a feedback is cloud cover."[14]

Veizer and Shaviv find little correlation between Earth's climate and CO_2 over the 500 million years. They note that CO_2 levels have been as much as eighteen times higher than today during the Veizer temperature record and were ten times higher than today's during the frigid Ordovician glacial period about 440 million years ago. This, to the researchers, suggests that "CO_2 is not likely to be the principal climate driver."[15]

The two scientists warn that the billion-dollar General Circulation computer models predicting dangerous global warming from CO_2 are particularly weak in modeling changes in the clouds—which may be the most important thermostat in the Earth's atmosphere.

In Veizer's 2005 paper for *Geosciences Canada*, he says, "Neither atmospheric carbon dioxide nor solar variability can alone explain the magnitude of the observed temperature increase over the last century of about 0.6 degrees Celsius." Therefore, an amplifier is required. "In the [global] climate models, the bulk of the calculated temperature increase is [assumed to be] positive water vapor feedback. In the sun-driven alternative, it may be the cosmic ray flux, energetic particles that hit the atmosphere, potentially generating cloud condensation nuclei. Clouds then cool, reflecting the solar energy back into space."[16]

How do we decide between the CO_2 theory and the sun–climate connection? Veizer says, "Cosmic rays . . . also generate the [so-called solar isotopes] such as beryllium-10, carbon-14, and chlorine-36. These can serve as indirect proxies for solar activity, and can be measured e.g., in ancient sediments, trees, and shells. Other proxies, such as oxygen and hydrogen isotopes, can reflect past temperatures, while stable carbon isotopes [reflect] levels of carbon dioxide."[17]

Veizer finds that "empirical observations on all time scales point to celestial phenomena as the principal driver of climate . . . with greenhouse gases acting only as potential amplifiers. . . . The tiny carbon cycle is piggybacking on the huge water cycle (clouds included), not driving it." He concludes, "Models and empirical observation are both indispensable tools of science, yet when discrepancies arise, observations should carry greater weight than theory."[18]

Veizer and Shaviv conclude that a doubling of today's CO_2 levels would only increase global temperatures a modest 0.75 degrees Celsius. The UN's IPCC, meanwhile, estimates an increase of 1.5 to 5.8 degrees Celsius from such a CO_2 increase, two to seven times as much. Some modelers have claimed that warming could increase by 11.5 degrees Celsius.[19]

The Veizer-Shaviv projection roughly matches the rate of warming observed by weather satellites in the lower atmosphere since 1979—which is well below the rate of increase seen in surface thermometers.

THE OCEAN–IRON FEEDBACK LOOP

Much of the world's ocean water is critically short of iron, a micronutrient needed by the huge masses of phytoplankton that make up the lowest rung on the marine food chain. Researchers have recently dumped iron filings into iron-poor waters—and triggered major surges in phytoplankton growth.

During the cold, dry ice ages, huge tracts of forest and prairie turn to desert. Iron dust from Patagonia, for instance, blows into the iron-starved waters of the Southern Ocean. Then a surge in phytoplankton growth pulls large amounts of CO_2 from the air. This feedback loop amplifies the global cooling trend.

The process operates too slowly to make much difference during the 1,500-year climate cycle. At the end of an ice age, however, some researchers estimate the ocean–iron feedback loop could account for half of the CO_2 released from the oceans into the atmosphere. This, in turn, amplifies the warming triggered by the sun's variation. The ocean–iron feedback loop endorses the reality that massive long-term processes on Earth have impacted the virtually continuous climate changes that have occurred throughout the Earth's history.

SUMMING UP THE SUN–CLIMATE EVIDENCE

The Armagh Observatory in Northern Ireland has weather observations going back to 1795. Its director, Richard Butler, says the records show that "it gets cooler when the sun's cycle is longer, and Armagh is warmer when the cycle is shorter."[20]

Richard Willson of Columbia University and the Goddard Institute says the sun's radiation has increased by 0.05 percent per decade since the late 1970s, which is as far back as we have high-precision satellite data to confirm the trend. Willson is not sure if the trend extends farther back in time, but "other studies suggest it does."

Svensmark's cosmic ray observations tie the formation of low clouds to the solar cycle and to a climate forcing four times as great as the variations in the sun's irradiance.

The Haigh and Shindell climate modeling seems to confirm ozone chemistry as a second factor in amplifying the sun's small variations in irradiance.

Veizer and Shaviv give us real-world isotopes to measure the impact of cosmic rays from the galaxy that correlate with physical evidence of the Earth's past climate changes—and also offer the scientific rule the actual measurements of solar-related isotopes are far more powerful evidence than computer models which have never been verified.

Against this real-world evidence, the Greenhouse Theory has only a warming trend that started rapidly and has now turned erratic. That's exactly the pattern of the natural, solar-driven warmings in previous 1,500-year cycles. The warming began too early and too suddenly for man-made CO_2 to be a likely candidate as its driving force. There is no overarching polar warming trend in either the Arctic or the Antarctic, as the Greenhouse Theory says there should be. There is little warming in the lower atmosphere, which the Greenhouse Theory says should be warming faster than the Earth's surface.

The strongest allies of the Greenhouse Theory are:

- Computer models that cannot explain past temperatures, let alone accurately forecast future ones, and whose funding depends on the public's fear of radical warming.
- Activists who oppose modern technology, abhor expanded human populations, and especially hate the low-cost energy that alleviates human poverty and misery. They say we must we renounce attractive lifestyles, give up high-yield farming, shorten millions of lives, and put more pressure on Third World forests for fuelwood.
- European politicians.
- Journalists looking for scary headlines.
- Various national and international bureaucracies and UN-appointed members and staff of the Intergovernmental Panel on Climate Change.

It's not a very impressive lineup.

NOTES

1. Henrik Svensmark, "Influence of Cosmic Rays on Earth's Climate," *Physical Review Letters* 81 (1999): 5027–30.

2. Alex Kirby, "Cosmic Rays 'Linked to Clouds,'" BBC News, 19 October 2002, <news.bbc.co.uk/2/hi/science/nature/2333133.stm>.

3. "NASA Study Finds Increasing Solar Trend that Can Change Climate," Goddard Space Flight Center "Top Story," press release, 20 March 2003.

4. Paal Brekke, solar physicist with the European Space Agency, "Viewpoint," BBC News, 16 November 2000.

5. M. Stuiver, "Variations in Radiocarbon Concentration and Sunspot Activity," *Journal of Geophysical Research* 66 (1962): 273–76.

6. C. A. Perry and K. J. Hsu, "Geophysical, Archaeological, and Historical Evidence Support a Solar-Output Model for Climate Change," *Proceedings of the National Academy of Sciences USA* 97 (2000): 12433–438.

7. D. T. Shindell et al., "Solar Forcing of Regional Climate Change during the Maunder Minimum," *Science* 294 (2001): 2149–152.

8. N. D. Marsh and H. Svensmark, "Low Cloud Properties Influenced by Cosmic Rays," *Physical Review Letters* 85 (2000): 5004–7.

9. Philip Ball, "Solar Blow to Low Cloud Could be Warming Planet," *Nature*, 6 December 2000.

10. H. Svensmark, "Influence of Cosmic Rays on Earth's Climate," Danish Meteorological Institute, *Physical Review Letters* 81 (1998): 5027–30.

11. J. D. Haigh, "The Effects of Change in Solar Ultra-Violet Emission on Climate," paper presented at the American Association for the Advancement of Science annual meeting, Philadelphia, February 1998.

12. Drew Slindell et al., "Solar Cycle Variability, Ozone, and Climate," *Science* 284 (9 April 1999): 305–8.

13. N. Shaviv and J. Veizer, "Celestial Driver of Phanerozoic Climate?" *Geological Society of America* 13 (2003): 4–10.

14. Shaviv and Veizer, "Celestial Driver of Phanerozoic Climate?" 4–10.

15. Shaviv and Veizer, "Celestial Driver of Phanerozoic Climate?" 4–10.

16. J. Veizer, "Celestial Climate Driver: A Perspective from Four Billion Years of the Carbon Cycle," *Geosciences Canada* 32, no. 1 (2005): 13–30.

17. Veizer, "Celestial Climate Driver," 13–30.

18. Veizer, "Celestial Climate Driver," 13–30.

19. Shaviv and Veizer, "Celestial Driver of Phanerozoic Climate?" 4–10.

20. "Sun's Warming Influence 'Under-estimated,'" BBC News, 28 November 2000, <newsbbcco.uk/hi/English/sci/tech/newsid_1045000/1045327.stm>.

Chapter Fourteen

The Baseless Fears

Millions of Human Deaths from Warming

Visions of Disasters:

"Global warming killed 150,000 people in 2000, and the death toll could double again in the next 30 years if current trends are not reversed, the World Health Organization said Thursday. . . . 'We see an approximate doubling in deaths and in the burden in healthy life years lost' by 2030, said WHO scientist Diarmid Campbell-Lendrum."[1]

"The world's leading authority on global warming, the Intergovernmental Panel on Climate Change, has concluded that unchecked global warming will cause a significant increase in human mortality due to extreme weather and infectious diseases. No country, even industrialized nations like the United States, will escape these impacts. . . . Malaria could become even more common in the U.S. as global warming worsens. IPCC scientists project that as warmer temperatures spread north and south from the tropics, and to higher elevations, malaria-carrying mosquitoes will spread with them. They conclude that global warming will likely put as much as 65 percent of the world's population at risk of infection."[2]

HEALTH RISKS FROM HEAT AND COLD WEIGHED

There are certainly deaths and illnesses due to heat waves. These typically include heat stroke, heart attacks, and asthma attacks. Deaths and hospitalizations from such heat waves make headlines whenever the temperature hits very high levels.

Yet we see the same sort of headlines during cold waves! The elderly die in inadequately heated homes. People get skull fractures from falls on the ice. Men die of heart attacks while shoveling snow. People get colds, flu, pneumonia, and other respiratory diseases. Infectious diseases proliferate. Hospital admissions rise.

The advocates of global warming present a fairly simplistic theory, that higher temperatures will drive more extreme weather events, and that these heat-related events will raise human death rates. But, overall, cold weather is much more effective at killing people than heat waves.

From 1979 to 1997, extreme cold killed roughly twice as many Americans as heat waves, according to Indur Goklany of the U.S. Department of the Interior.[3] Cold spells, in other words, are twice as dangerous to our health as hot weather.

Heat is becoming a less and less important factor in human health as air conditioning spreads. Heat-related mortality in twenty-eight major U.S. cities from 1964 through 1998 dropped from 41 per day in the 1960s to only 10.5 per day in the 1990s.[4] A large cohort study comparing households with and without air conditioning in the early 1980s found a 41 percent lower death rate for the air-conditioned households during hot months.

In Germany, heat waves were found to *reduce* overall mortality rates slightly, while cold spells led to a significant increase in deaths![5] The German authors say that the longer a cold spell lasts, the more pronounced the excess mortality—and the higher death rates seem to persist for weeks.

Hot spells, in contrast, cause a short surge in deaths—followed by a period of *lower* death rates that persists for more than two weeks.

Global warming would raise maximum summer temperatures modestly while it would raise winter minimum temperatures significantly. *Both factors should help slightly reduce human death rates*!

Cardiovascular Episodes

A high proportion of modern deaths are due to cardiovascular problems, and cold weather is far more dangerous for people with heart problems. In cold weather, the body automatically constricts blood vessels to conserve body heat. That raises blood pressure and doubles the heart attack risks of people with high blood pressure. People with hypertension are also at increased risk because they have overreactive blood vessels and their blood vessels constrict even more than the average person's when the temperatures drop suddenly.

A study of Siberian health records between 1982 and 1993, where stroke rates are among the highest in the world, found a 32 percent higher stroke risk on cold days than on warm ones.[6] Korean evidence shows much the same pattern. Cold periods were associated with higher risks of strokes, "with the strongest effect being seen on the day after exposure to cold weather."[7]

Respiratory Diseases

Numerous studies from around the world show a connection between cold weather and respiratory diseases. A Norwegian study found 47 percent more respiratory deaths in winter than in summer.[8] In London, a 1 degree Celsius drop in mean temperature (below 5°C) was associated with a 10.5 percent increase in all respiratory disease consultations.[8] In Brazil, the adult death rate changes due to a 1 degree Celsius cooling were twice as great as death rate changes due to a similar warming—and 2.8 times greater among the elderly.[10]

In the United States, temperature variability is the most important element of climate change for respiratory deaths, though the reasons for this do not seem to be clear to physicians.[11] The risk from temperature variability is important to our analysis, nonetheless, because a fifty-year study of 1,000 U.S. weather stations found *temperature variability declines very substantially with climate warming*.[12] This means that the benefit of warming should extend throughout the entire year.

Heat-Related Deaths from All Diseases

In London, low temperatures made the only significant weather-related impact on both immediate deaths (one day after the temperature extreme) and medium-term mortality (up to twenty-four days after the extreme).[13]

Closely allied studies looked at mortality of the elderly (sixty-five to seventy-four years) in north Finland, south Finland, southwest Germany, the Netherlands, Greater London, northern Italy, and Athens, Greece.[14] They found *annual cold-related deaths were nearly ten times greater than annual heat-related deaths*.

MALARIA AND VECTOR-BORNE DISEASES

Activists of Man-Made Warming Threaten More Malaria:

"The combined effects of increased warmth and the greater volume of standing water brought by storms create malaria epidemics by providing breeding sites and a speeded-up life cycle. In Africa, where the death toll from malaria is highest, mosquitoes carrying the disease are spreading into mountain areas previously too cool for them to thrive."[15]

"In the early 1930s, 36 percent of Southerners, black and white, had hook-worm infestations. . . . In addition to hookworm infestations, nearly a third of the inhabitants of the rural South had chronic malaria from living in houses with no window screens. As if malaria and hookworm were not enough, add to those the scourge of malnutrition. . . . About 25 percent of the Southern population had pellagra, a disease caused by lack of tryptophan, an essential amino acid. . . . The diet of Southern tenant farmers then consisted almost en-tirely of corn bread and hog fat belly. . . . Through the efforts of the Rocke-feller Foundation and the public health departments . . . a massive education campaign was set into motion. The final result . . . was to render the South free of all three of its ravaging diseases: malaria, by controlling mosquito re-production [with DDT] and screening doors and windows; pellagra, by teach-ing the need for meat in the diet, and by vitamin supplements; and hookworm, by the wearing of shoes. I don't know of a more dramatic and successful story . . . [to] eliminate disease on such a wide scale."[16]

Why Washington Politicians Don't Need Mosquito Netting

Advocates of global warming have warned of increased malaria infections and deaths as global temperatures increase. However, this ignores the reality that malaria epidemics have occurred as far north as the Arctic Circle. The worst known outbreak was in Russia during the 1920s, with 16 million sick and 600,000 deaths.

If air temperatures alone were the key factor, mosquito-borne diseases, in-cluding malaria and yellow fever, would already have become major threats to Americans again as we flocked to the Sun Belt and water-shore environ-ments during the air-conditioning era. Instead, because of modern medicine and technology, we have found ourselves free to enjoy seashores, riverbanks, southern living, and marsh-side homes with fewer health worries than any people in history.

Let us quote Dr. Paul Reiter from the Center for Disease and Prevention (CDC) journal *Emerging Infectious Diseases*:

> Discussions of the potential effects of the weather include predictions that malaria will emerge from the tropics and become established in Europe and North America. The complex ecology and transmission dynamics of the dis-eases, as well as accounts of its early history, *refute such predictions*. Until the second half of the 20th century, malaria was endemic and widespread in many temperate regions, with major epidemics as far north as the Arctic Circle. From 1564 to 1730—the coldest period of the Little Ice Age—malaria was an impor-tant cause of illness and death in several parts of England. Transmission began

to decline only in the 19th century, when the present warming trend was well under way.[17]

Reiter notes that Chaucer wrote about malaria in the fourteenth century and Shakespeare mentioned malaria (which he called ague) in eight of his plays — in a nontropical region during the Little Ice Age.

In medieval times, the only "treatments" for malaria were alcohol and opium. Only in the late seventeenth century did Europe begin to learn about quinine as an effective treatment for the disease after its discovery in the 1500s by the Spanish conquistadors among Andean Indians.

Malaria *prevention* became possible only in the latter half of the twentieth century. We finally developed pesticides that could kill the malaria-carrying mosquitoes, and screen doors and windows became cheaply available to keep mosquitoes out of homes, schools, and workplaces. According to Dr. Reiter:

> It was not until the advent of DDT, after World War II, that a concerted attempt could be made to eradicate the disease from the entire [European] continent. At the same time the Communicable Disease Center was set up in Atlanta to eliminate malaria from the United States, where it was still endemic in 36 States, including Washington, Oregon, Idaho, Montana, North Dakota, Minnesota, Wisconsin, Iowa, Illinois, Michigan, Indiana, Ohio, New York, Pennsylvania, and New Jersey. (!!!)

Thanks in large part to DDT, England was then freed of malaria during the late 1950s and Europe was declared malaria-free in 1975. By 1977, 83 percent of the world's population lived in regions where malaria had been eradicated or control activities were in progress.

LIFE EXPECTANCY IN A HOTTER WORLD AFTER 2050

Rockefeller University demographer Shiro Horiuchi has written a remarkable summation of twenty-first-century life expectancies. He notes that the average lifespan of early humans was about twenty years, and that this has been extended in today's affluent countries to about eighty years.

The first wave of major life-lengthening achievements occurred in the second half of the nineteenth century and the first half of the twentieth, as modern medicine radically reduced deaths from infectious and parasitic diseases, poor nutrition, and risks from pregnancy and childbirth. In the second half of the twentieth century, "mortality from degenerative diseases, most notably heart disease and stroke, started to fall."[18]

The reduction in heart disease and stroke deaths was, of course, most notable among the elderly, says Horiuchi. Some critics have suspected that these mortality declines have been achieved "through postponing the deaths of seriously ill people," he notes. However, he says the data from the United States demonstrate that "the health of the elderly greatly improved in the 1980s and 1990s, suggesting that the extended length of life in old age is mainly due to better health, rather than prolonged survival in sickness. . . . These days, the existence of a biological limit to human longevity is considered questionable."[19]

Nothing in any global warming scenario is likely to alter the expected results of modern medicine and health care: better health and longer lives in the twenty-first century, whether the century's higher temperatures are natural or man-made. Indeed, the outlook is that more and more of the world's human population is likely to live longer and healthier lives—unless the fear of man-made global warming leads human societies to restrict or abandon the gains in economic prosperity achieved through abundant, low-cost energy, and further advances in science and technology.

NOTES

1. "Global Warming Killing Thousands," Reuters, 11 December 2003, <www.wired.com/news/politics/0,1283,62737,00.html?> (1 October 2005).

2. Sierra Club, "Global Warming Impacts: Health Effects," 2004, <www.sierraclub.org/globalwarming/health/conclusions.asp>.

3. I. M. Goklany and S. R. Straja, "U.S. Trends in Crude Death Rates Due to Extreme Heat and Cold Ascribed to Weather, 1979-1997," *Technology* 7S (2000): 165–73.

4. R. E. Davis et al., "Changing Heat-Related Mortality in the United States," *Environmental Health Perspectives* 111 (2000): 1712–718.

5. G. Laschewski and G. Jendritzky, "Effects of the Thermal Environment on Human Health: An Investigation of 30 Years of Daily Mortality Data from SW Germany," *Climate Research* 21 (2002): 91–103.

6. V. L. Feigin et al., "A Population-Based Study of the Associations of Stroke Occurrence with Weather Parameters in Siberia, Russia (1982-1992)," *European Journal of Neurology* 7 (2000): 171–78.

7. Y.-C. Hong et al., "Ischemic Stroke Associated with Decrease in Temperature," *Epidemiology* 14 (2003): 473–78.

8. P. Nafsted, A. Skrondal, and E. Bjertness, "Mortality and Temperature in Oslo, Norway, 1990–1995," www.ncbi.nlm.nih.gov/entrez.

9. S. Hajat and A. Haines, "Associations of Cold Temperatures with GP Consultations for Respiratory and Cardiovascular Disease amongst the Elderly in London," *International Journal of Epidemiology* 31 (2002): 825–30.

10. N. Gouveia et al., "Socioeconomic Differentials in the Temperature-Mortality Relationship in Sao Paulo, Brazil," *International Journal of Epidemiology* 32 (2003): 390–97.

11. A. L. F. Braga and J. Schwartz, "The Effect of Weather on Respiratory and Cardiovascular Deaths in 12 U.S. Cities," *Environmental Health Perspectives* 11 (2002): 859–63.

12. S. M. Robeson, "Relationships between Mean and Standard Deviation of Air Temperature: Implications for Global Warming," *Climate Research* 22 (2002): 205–21.

13. W. R. Keatinge and G. C. Donaldson, "Mortality Related to Cold and Air Pollution in London after Allowance for Effects of Associated Weather Patterns," *Environmental Research* 86A (2001): 209–16.

14. W. R. Keatinge et al., "Heat Related Mortality in Warm and Cold Regions of Europe: Observational Study," *British Medical Journal* 321 (2000): 670–73.

15. Paul Brown, "Global Warming Kills 150,000 A Year," The London Guardian, 12 December 2003.

16. Clifton Meador, Med School (Franklin, TN: Hillsboro Press, 2003), 67–82.

17. Paul Reiter, "From Shakespeare to Defoe: Malaria in England in the Little Ice Age," *Emerging Infectious Diseases* 6 (January–February 2000): 1–11.

18. S. Horiuchi, "Greater Lifetime Expectations," *Nature* 405 (15 June 2000): 789–92.

19. Horiuchi, "Greater Lifetime Expectations," 789–92.

Chapter Fifteen

Powering the Future

Can We Depend on Renewable Energy?

The Wind Blows Freely:

"On a clear day, you can see the boats on the horizon off Cape Cod, their sails taut against the storied wind that draws serious yachtsmen. . . . It is this Massachusetts wind that developer Jim Gordon and his company, Cape Wind Associates, wish to harness. They hope to build the nation's first big offshore commercial wind farm, with 130 giant turbines anchored on Horseshoe Shoal. The $700 million project, with 426-foot turbines spaced six to nine football fields apart over a 24-mile area, could produce enough electricity to power most of Cape Cod. . . . But an intense squabble over wind, views, fish, birds, tourism, and the right to build in this water paradise has resulted in odd alliances. Homeowners and business owners say they and thousands of tourists want to look out on water and sails, not turbines."[1]

"Anger Over Fuel Taxes Spills Across Borders. . . . The trouble began in France . . . but soon, people outraged by rising fuel prices seemed to be venting their feelings everywhere. They shut down the center of Brussels, the EU capital, nearly paralyzed Britain, and are continuing disruptions in Spain and Germany. This week, in Spain, fishermen used their boats to seal off ports, and farmers blockaded fuel distribution centers. . . . What unites European protesters is not anger at oil-producing countries whose oil prices are at 10-year highs. Instead, their anger is directed at their own governments' taxes . . . which account for up to 76 percent of the price."[2]

Sweden Reconsiders Nuclear Power:

"On April 4, 2004, Sweden's Liberal Party (Folkpartiet), currently in opposition, announced a new nuclear policy. They now want to reverse the decisions of a 1980 referendum to phase out nuclear reactors. Said Jan Bjorklund, vice chairman of the Liberal Party, 'In l980, the voters were told sun and wind would replace nuclear power. Now we see that oil and gas are the realistic alternatives. This is not acceptable.' The main reason given for the new policy is concern for global warming, as Sweden is about to increase its output of carbon dioxide with new fossil-fuel power plants. The party also points out that more than 50 percent of Sweden's current population is too young to have voted in 1980."[3]

Windmills are huge structures that produce very little electricity—and then only when the wind blows within certain speed ranges. It would take nearly two thousand new 750-kilowatt wind turbines operating at the normal 28 percent capacity factor to produce as much electricity as one 500-megawatt gas-fired combined cycle base-load generating plant.

Electricity from wind energy does not much reduce emissions from coal and other fossil-fueled electric generating plants. Because wind turbines produce only intermittently, other generating plants have to be immediately available—either running at less than full capacity or in "spinning reserve"—to supply electricity when wind speed drops or disappears. The backup plants still produce emissions while in this backstopping mode.

North Dakota and other remote states have significant wind resources, but windmills do not produce enough electricity to justify the cost of building new transmission lines.

Wind energy is growing rapidly in percentage terms only because the base from which it starts is tiny. EIA projects that wind energy will supply just 0.0025 percent of the U.S. electricity generation in 2020.

The cost of wind energy is not competitive and wind farms would not be constructed without massive subsidies. Developers are building "wind farms" for four reasons—all of which shift costs from wind developers and hide them in unsuspecting Americans' tax bills and monthly electric bills:

1. Lucrative tax shelters,
2. State Mandates,
3. "Greenwashing" for public relation purposes,
4. Green pricing programs.

The Department of Energy (DOE) and its predecessor agencies have spent hundreds of millions of tax dollars on wind energy research and development (R&D). However, most of the wind turbines being used in the United States are from Danish companies.

Windmills can be an economical source of electricity in remote areas where electricity is otherwise unavailable or where costs of building electricity distribution lines are prohibitive. Even then, windmills produce only when the wind is blowing and the feasibility of storing electricity in batteries is limited and very expensive. Windmills work well for pumping water—which can be stored. That is how and why the famous old windmills on American farms were built, used, and became famous.

SOLAR POWER SOLUTIONS

Comments:

"A major European chip maker said this week it had discovered new ways to produce solar cells which will generate electricity twenty times cheaper than today's solar panels. . . . Most of today's solar cells, which convert sunlight into electricity, are produced with expensive silicon. . . . The French-Italian company expects cheaper organic materials such as plastics to bring down the price of producing energy. Over a typical 20-year life span of a solar cell, a single produced watt should cost as little as $0.20, compared with the current $4. The new solar cells would even be able to compete with electricity generated by burning fossil fuels such as oil and gas, which costs about $0.40."[4]

"In addition, at least 15 states now use 'public benefits funds' to subsidize renewable energy. . . . Gail Stocks' husband, Ian, says his family's 2.5 kilowatt solar-panel system cost $21,000, but their out-of-pocket cost was only $9,000. It cuts their electric bill by a third."[5]

"However, as an energy source, solar radiation is relatively dilute. Impressive amounts of (desert) land area would have to be devoted to this use in order to replace fossil fuel supplies. . . . [C]omplete replacement . . . in the United States would require total collector fields on the order of 50,000 square miles, about 1 percent of the total U.S. land area (Ogden and Williams, 1989) . . . obtaining the same power from biomass grown on energy farms would require more than 10 times that area."[6]

WHAT HAPPENS WHEN THERE ISN'T ENOUGH ELECTRICITY?

To judge the importance of the cutbacks in energy use demanded by the Kyoto treaty, we need only look at California's rolling blackouts and the statewide crisis they caused back in 2001. California's electrical generating capacity had not kept up with its rising demand. Because of environmental opposition to new coal, oil, gas, and nuclear power plants the state had not built any new power plants in more than twelve years. Then, in 2000, drought cut the supplies of electricity from western hydroelectric dams. Finally, natural gas prices soared as the whole Western world tried to meet politically driven "clean air" demands by massively substituting natural gas for coal and oil—all at once.

The shortage of electricity hit California hard. "Interruptible" customers were cut off first. But then rolling blackouts hit 1.5 million regular customers, blacking out the schools and traffic lights, crashing the computers, and stalling the elevators. The economic losses to the state's economy were huge. Cut-off irrigation pumps let thirsty crops die. In industry, the blackout shut down the "clean rooms" producing the high-tech electronics that had been leading the state's economic growth. Major employers announced their intention to leave the Golden State. The city of San Francisco sued the power companies, charging that they were taking advantage of a deregulated market to profiteer.

Outsiders pointed instead at California's ludicrous deregulation scheme, which had forced utilities to sell their generating plants and buy their power on the spot market. The companies had been barred from signing long-term supply contracts—even though they were being forced to sell at regulated low prices "to protect consumers." It was an ideal recipe for bankruptcy. And, sure enough, California's power companies went bankrupt in short order.

Then the state intervened. It quickly spent its projected $8 billion revenue surplus, and more, to provide the electricity that no one had signed long-term contracts to deliver.

In the years since the rolling blackouts ended, California has been burdened with surpluses of electricity contracted for at the very high emergency prices. In 2004 alone, the state was expected to resell 25 percent of its contracted power as "surplus," at a loss of another $772 million. Consumer power rates have meanwhile risen by 15 to 50 percent.

What would have happened if the whole United States, or the entire First World, had been caught like California, without adequate power to supply homes, schools, and businesses?

Man-made warming advocate and biologist Stephen Schneider has denounced the idea that scientific discoveries might enable humanity to protect the environment and still keep high human standards of living. Schneider calls such research "an irresponsible palliative." He says, "It evades the need for a real cure, such as curbing the consumption of the rich and the population growth of the poor, and charging polluters for their use of the atmosphere as a free sewer." Schneider is also hard at work on behalf of a "world government" that would have the authority to "enforce responsible use of the global commons."[1]

The population growth of the poor has essentially been curbed already, however, and with impressive speed. Births per woman in the Third World have dropped from about 6.2 in 1960 to about 2.8 today—and stability is 2.1. Thus, the poor nations have already come 75 percent of the way to population stability. The "rich" nations are having 1.7 children per woman or less. The UN Population Division expects world population to peak about 2035 or 2040, at something between 8 and 9 billion. After that, human numbers will slowly decline.

Schneider says he is worried about progress in "curbing the consumption of the rich."[8] However, he seems not to realize that richer people have fewer children, feed themselves from much less land per capita, gradually but hugely reduce industrial pollution, generally plant more trees than they cut, do most of the research on environmental preservation, and make most of the investments in actual conservation. Countries with per capita incomes above $8,000 are the hope for sustainability, not the curse of the planet.

Primitive peoples have never "lived in harmony with nature" as the urban legend would have us believe. They have, instead, exploited nature unsustainably because their populations always crept upward to the absolute limits of their resources—and then beyond.[9] More people were an advantage in war, and victory in war increased the tribe's resources.

Today, poor peoples practice slash-and-burn farming, cook and heat with wood from trees they don't replant, and kill endangered animals for supposed aphrodisiacs.

The most vivid recent example of tribal competition for resources was in Rwanda in 1994. That's when Hutu tribespeople slaughtered nearly one million of the Tutsis who lived among them—mostly up close and personally, with machetes. The Hutus apparently feared there would not be enough resources in the increasingly crowded Central African highlands for both tribes. Rwanda did have an agricultural experiment station, but it had not achieved enough yield gains for Rwanda's corn and potato crops to reassure both tribes about their common future.

ENERGY TECHNOLOGY FOR A GLOBAL FUTURE

For a look at the world's possible energy future, we turn to real experts in that field: the authors of "Advanced Technology Paths to Global Climate Stability," *Science*'s "Compass" feature on 1 November 2002. The authors of this article are staff members from such institutions as the U.S. government's Lawrence Livermore Laboratory in California, the Physics Department at the University of California–Irvine, the Institute of Space Systems Operations at the University of Houston, the MIT Laboratory for Energy and the Environment, the Exxon Mobil Research and Engineering Company, the Plasma Physics Division of the U.S. Naval Research Laboratory, and NASA. The lead author was Martin L. Hoffert, a New York University physicist. The team's basic conclusion was that cutting CO_2 levels would "require Herculean effort." They say CO_2 is a key element of modern society that "cannot be regulated away."[10]

The IPCC said in its Third Assessment (2001) that "known technological options could achieve a broad range of atmospheric CO_2 stabilization levels, such as 550 ppm, 450 ppm or below over the next 100 years or more. . . . Known technological options refer to technologies that exist in operation or pilot plant stage today. It does not include any new technologies that will require drastic technological breakthroughs."[11]

The Hoffert team differs radically, "This statement does not recognize the CO_2 emission-free power requirements implied by the IPCC's own reports and is not supported by our assessment."[12] That says "No!" as loudly as the decorum of a scientific journal will permit.

Global power consumption today is about 12 trillion watt-hours per year, of which 85 percent is fossil fueled. By 2052, the world will demand an additional 10 to 30 trillion watt-hours per year. We would have to achieve that emission-free increase even as we shifted from a system that today produces very little emission-free energy. In fact, the small amount of emission-free energy we do produce is mostly from hydroelectric dams and nuclear power plants—which are strongly opposed by the same green movement that says all our current warming is due to human greenhouse emissions. In other words, the IPCC has made a political statement about producing emission-free energy that is totally out of line with energy realities.

The Nonanswer of Renewables

The Hoffert team says the efficiencies of "renewable" energy sources are very low and they have low "power densities."[13] In other words, virtually all of the energy sources recommended by the Greens would require huge amounts of land, the scarcest of all resources. As an example, a proposed set of 1,300 wind turbines in North Dakota would have a capacity of only 2,000 megawatts. The proposed 1,300 wind turbines would, however, require fifty to one hundred acres per turbine, for a total of some one hundred to two hundred square miles of land. On a good day, they would produce about 2 percent of the increase needed in U.S. generating capacity this year—and only half as much electricity as one large nuclear power plant.

The controversial wind farm proposed for Cape Cod would not even provide all of the electricity needed for the relatively small population of that posh resort.

Suppose we try to produce from renewable sources just the additional 30 trillion watt-hours per year of electricity that is likely to be demanded for the year 2050 from a combination of biofuels (33 percent), solar (33 percent), and wind (33 percent). The Hoffert team is suggesting that 10 trillion watt-hours per year of biofuels would require 15 million square kilometers of cropland to be converted from wildlife habitat. Since the biofuel land will be of poorer quality than current cropland, and organic farming methods will be mandated, the biocrops might need up to 30 million square kilometers. Solar panels to produce another 10 TW of electricity would require another 220,000 square kilometers or so of wildlands for photovoltaic arrays plus the land for their associated transmission lines, service roads, maintenance yards, and so on. Ten trillion watt-hours per year of wind energy will need windmills occupying more than 600,000 square kilometers. Much of the land under the windmills could be farmed, but it could not be forest, the best wildlife habitat. Logically, some of the windmills could also be offshore, but the ongoing controversy about coastal views does not look promising for offshore wind power.

All told, the "green power" land requirement would probably clear forests equal to the land area of South America (22 million square kilometers) and could require the land area of China (10 million square km) and India (3 million square km) as well. Is this what anyone could call conserving Nature? And, we would still have to build thousands of additional power plants for the cloudy, windless, and/or high-wind days. And, to step in for biofuels in the case of any biocrop pest, disease, or weather failures in our biofuel crops.

The "renewable" energy sources have been politically useful as imagined alternatives to the power plants we could see and deplore. As we actually try to build the wind farms and big solar arrays, the very people who demanded them shift into opposition—because there never seems to be a "right place" for them. If renewables are to make an important contribution, however, there will have to be a great many "right places."

In its heart, the green movement has long believed that First World people are too rich and should live more simply. The Greens probably assume that even if they're wrong about renewable energy, they'll at least be pushing us in the right direction—toward much lower standards of living. They probably have not given great weight to the broader life-span and quality-of-life implications of a human shift back to low-tech living.

Doubling Transport Fuel Efficiency

The *Science* "Compass" authors say that it should be possible to double the fuel efficiency of current world vehicles, with or without banning sport-utility vehicles. This would probably have little to do with the electric cars which activists demanded for California—and which Californians, quite rightly, refused to buy. The electrics' performance and load-carrying capacities were terrible and their operating range before recharging was very short.

On the other hand, the Toyota hybrid Prius was named the 2004 Car of the Year by the performance-oriented *Car and Driver*. Its editors said it combined surprising performance and useful size with outstanding fuel economy (60 mpg in the city and 43 mpg on the highway). It seems likely that hybrid vehicles will play a strong role in advancing vehicle fuel mileage in a wide variety of cars, light trucks, and—yes!—SUVs.

"Unfortunately," the Hoffert authors warn, "the effects of such efficiency could be overwhelmed if China and India follow the U.S. path from bicycles and mass transit to cars. (Asia already accounts for more than 80 percent of world petroleum consumption growth.) As a result, carbon-neutral fuels or CO_2 'air capture' may be the best alternatives to develop for vehicles. . . . The simplest air capture is forestation . . . but trees' capacity to absorb CO_2 is limited."[14]

Hydrogen may not be much help, the Hoffert team warns: "Per unit of heat generated, more CO_2 is produced by making H_2 from fossil fuel than by burning the fossil fuel directly."[15] Emission-free H_2 manufactured by water electrolysis powered by renewable or nuclear sources is not yet cost effective. Hydrogen, at this point, is not so much a power source as it is a transmission medium for energy.

The Hoffert team thinks "clean" coal technologies will be important, since so much coal is available at relatively low cost. "Coal and/or biomass and waste materials are gasified in an oxygen-blown gasifier, and the product is cleaned of sulfur and reacted with steam to form H_2 and carbon monoxide. After heat extraction, the carbon monoxide is turned into carbon dioxide, which can be sequestered, and the H_2 can be used for transportation or electricity generation."[16] The key problem is still how to get rid of the CO_2.

The Hoffert authors are not terribly optimistic about Schneider's favorite idea of recapturing carbon and sinking it into the earth or the oceans. Unless other emission-free primary power sources are available by mid-century, "enormous sequestration rates could be needed to stabilize atmospheric CO_2."[17] Very high levels of CO_2 in the oceans would make them more acid— changing ocean ecology.

Nuclear power also has some very real limitations. The Hoffert team points out that there is a uranium fuel shortage for the longer term, since 10 trillion watt-hours of energy per year would use up the current supply of recoverable uranium ores in less than thirty years.

The Hoffert team of energy experts clearly pins its hopes on still more technology, which may not make Stephen Schneider and his Green colleagues very happy. Specifically, they want nuclear fission and breeder reactors, followed ultimately by fusion reactors.

"Commercial [breeder reactors are] illegal today in the United States because of concerns over waste and proliferation (France, Germany, and Japan have also abandoned their breeding programs). Breeding could be more acceptable with safer fuel cycles and transmutation of high-level wastes to benign products. . . . Both fission and fusion are unlikely to play significant roles in climate stabilization without aggressive research and, in the case of fission, without the resolution of outstanding issues of high-level waste disposal and weapons proliferation."[18]

"Despite enormous hurdles, the most promising long-term nuclear power source is still fusion" say Hoffert and his team.[19] Their statement that this is our best bet simply underscores their assertion that it would take Herculean efforts or massive reductions in living standards to prevent continuing increases in humanity's CO_2 emissions.

Most of today's nuclear reactors are the light-water design pioneered for submarines in the 1950s. "Loss-of-coolant" accidents such as Three Mile Island could be avoided in the future by "passively-safe" reactor designs.

The Hoffert team may be unduly pessimistic about the long-term prospects of fission. Plentiful thorium is another candidate for fission reactors once uranium becomes too costly. Other experts consider that with the reprocessing of spent fuel and the use of breeder reactors, uranium fission reactors would be effective for thousands of years.

A political consensus on spent nuclear fuel disposal must also be achieved if more nuclear plants are to be built. So far, the environmental groups have led the opposition to centralized nuclear waste disposal sites such as Yucca Mountain in Nevada, though it seems likely that the cries of danger about Yucca Mountain are just a politically effective way to block another non-CO_2 energy source.

These positions suggest that the environmentalists don't take very seriously their own claims that global warming makes all other global dangers pale by comparison.

The Illusion of a Hydrogen Economy

Two of the most outspoken global warming activists, Amory Lovins and Jeremy Rifkin, have been proposing a national fleet of "Hypercars" powered by hydrogen which they say could not only provide our transportation, but when parked—"96 percent of the time"—could be plugged into a national power grid where they would produce two-plus kilowatts of electricity per vehicle. However, the "distributed energy generation network" touted by Lovins and Rifkin actually produces no energy of its own, and wastes two-thirds of the electricity generated by the original power plant turning the power first into hydrogen and then back into electricity. The "hydrogen economy" is an illusion.

ADAPTIVE TECHNOLOGY CHOICES

One American, asked to name the greatest invention of the twentieth century, answered "insulation." He'd grown up in a poorly insulated nineteenth-century mountain ranch house and it was his job to cut the family's firewood. Insulation is cheap, safe, sustainable, and helps reduce energy requirements whether the climate is warming or cooling. It is a fine "no-regrets" technology of the "install-it-and-forget-it" variety. More insulation in our buildings is a good idea whether the climate changes much or not.

Nuclear power plants are a cost-effective adaptive strategy for either a planet suffering from man-made climate change or a naturally warming world that will eventually run short of fossil fuels. Nuclear energy is the world's biggest source of non-fossil energy and emits virtually no greenhouse gases. The First World has generally turned away from nuclear power as activist lawsuits have driven up regulatory costs.

The Soviet Union's inexplicable decision to build its Chernobyl plant without a reactor containment structure also cast a pall over the whole nuclear industry for some years, though the Chernobyl site has now become a tourist attraction. Its death toll stands at less than fifty, with no surge of epidemic fatalities in sight.

Nevertheless, the world gets about 16 percent of its electricity from nuclear power, and France gets about three-fourths of its power from economical, standardized nuclear reactors. At this moment, Finland is building the only new nuclear plant in Europe. China, India, Japan, and South Korea are building sixteen new reactors, and China and India both say they will quadruple their nuclear power in the coming years.

There is an amazing array of energy sources that the eco-activists reject. Many of them are now reneging on wind power. Are there any major energy sources that the eco-activists *will* approve?

Is There Any Power Source the Greens Will Accept?

"Environmental groups have sued the federal government over geothermal projects it has approved in the remote Medicine Lake Highlands region considered sacred by Indian tribes. The suit, filed Tuesday and announced Wednesday, challenges approval of the first two geothermal power plants proposed by Calpine Corp. Both would be built within the Medicine Lake caldera, the remnant of an ancient volcano 30 miles east of Mt. Shasta . . . in northeastern California. The four environmental coalitions that filed the suit in Sacramento federal court contend the power projects . . . would turn an otherwise scenic natural area into 'an ugly, noisy, stinking industrial wasteland.'" The proposed geothermal power plants would emit no greenhouse gases nor would they produce radioactive waste. They would even feed into the existing Bonneville power grid without the need to create extensive new transmission lines and rights of way. And even this is unacceptable because it would create a few nine-story power plants on fifteen-acre pads in a remote location where almost no one would have to look at them.[20]

NOTES

1. Stephen Koff, Newhouse News Service, 29 September 2003, <www.newhousenews.com/archive/koff092903.html>.

2. Peter Finn, "Europe Unites in Gas Protests," *Washington Post*, 22 September 2000.

3. "That Was the Week that Was," Science and Environmental Policy Project, 10 April 2004.

4. "Discovery May Spur Cheap Solar Power," 2 October 2003, <www.CNN.com/technology>.

5. Mark Clayton, "Solar Power Hits Suburbia," *Christian Science Monitor*, 12 February 2004.

6. *Policy Implication of Greenhouse Warming: Mitigation, Adaptation and the Science Base, Panel on Implications of Greenhouse Warming* (Washington, D.C.: National Academy Press: 1992), 775.

7. S. Schneider, "Earth Systems Engineering and Management," *Science* 291 (2001): 417–21.

8. Schneider, "Earth Systems Engineering and Management," 417–21.

9. Stephen LeBlanc, "Constant Battles," *Archeology* (May/June 2003): 18–25.

10. M. Hoffert et al., "Advanced Technology Paths to Climate Stability: Energy for a Greenhouse Planet," *Science* 298 (1 November 2002): 981–87.

11. Summary for Policymakers, Working Group III, *IPCC Third Assessment Review, 2001*: 8.

12. Hoffert et al., "Advanced Technology Paths to Climate Stability," 981–87.

13. Hoffert et al., "Advanced Technology Paths to Climate Stability," 981–87.

14. Hoffert et al., "Advanced Technology Paths to Climate Stability," 981–87.

15. Hoffert et al., "Advanced Technology Paths to Climate Stability," 981–87.

16. Hoffert et al., "Advanced Technology Paths to Climate Stability," 981–87.

17. Hoffert et al., "Advanced Technology Paths to Climate Stability," 981–87.

18. Hoffert et al., "Advanced Technology Paths to Climate Stability," 981–87.

19. Hoffert et al., "Advanced Technology Paths to Climate Stability," 981–87.

20. Thompson, Don, "Environmentalists Sue over Medicine Lake Geothermal Plans," Associated Press, 20 May 2004.

Chapter Sixteen

The Ultimate Failure of the Kyoto Protocol

The Kyoto Protocol finally took effect on 16 February 2005, more than seven years after it was negotiated. At first glance, this seemed to be the moment of triumph for the environmental movement and the United Nations' Intergovernmental Panel on Climate Change. After seven years in limbo, the international treaty aimed at reducing humanity's CO_2 emissions finally came into force.

Kyoto's Obituary?

All was gloom, however, at the 10th Conference of Parties (COP 10) in Buenos Aires just prior to the installation proceedings in late 2004. At least one reporter covering the COP 10 meeting declared Kyoto's ultimate failure:

> The Kyoto Protocol is dead—there will be no further global treaties that set binding limits on the emissions of greenhouse gases after Kyoto runs out in 2012. . . . The conventional wisdom, that it's the United States against the rest of the world in climate change diplomacy, has been turned on its head. Instead, it turns out that it is the Europeans who are isolated. China, India, and most of the rest of the developing countries have joined forces with the United States to completely reject the idea of future binding greenhouse gas emission limits. At the conference here in Buenos Aires, Italy shocked its fellow European Union members when it called for an end to the Kyoto Protocol in 2012. These countries recognize that stringent emission limits would be huge barriers to their economic growth and future development.[1]

(Technically, Kyoto will continue to exist after 2012. However, the only emissions reduction that its member countries have agreed upon, 5.2 percent from 1990 levels, runs out.)

U. S. REJECTION OF KYOTO

Americans had at least four reasons to be less enthusiastic than West Europeans about the Kyoto Protocol.

First, America's two-party system and winner-take-all presidential elections leave the environmental movement on the fringes of the U.S. power structure. West European governments are almost always coalitions, and even minority parties like the Greens can gain important political leverage if they can deliver their voting blocs consistently at the polls.

Second, the United States has traditionally had fairly low fuel taxes due in part to America's big spaces and broadly spread economy.

Third, the United States regards its low fuel costs as an economic growth advantage, generating more and better jobs and more attractive suburban lifestyles. Europe regards itself as an urban society, and has historically favored its inner cities and subsidized rail lines over highway transport. Its politicians looked forward to saddling American businesses with energy taxes as high as Europe's. They hoped this would ease the political pressures on the European welfare states that were suffering mounting unemployment, Third World export gains, and the United States's much-higher recent economic growth rates. Western Europe was overjoyed when Al Gore and the Clinton administration signed the American economy up for Kyoto's energy constraints—and terribly disappointed when George W. Bush erased that U.S. commitment.

Fourth, Kyoto favored several important European countries when it selected 1990 as the base year from which emission reductions would be measured. Great Britain got Kyoto credit for closing down most of its antique, money-losing coal mines and shifting its industries to cleaner-burning natural gas from its North Sea wells. Germany got credit for clearing up the pollution-spewing, energy-inefficient industries of the old East German government. France was largely indifferent to Kyoto, depending heavily on its many standardized nuclear power plants.

When President George W. Bush took office, he called the Kyoto treaty "fatally flawed in fundamental ways," and his administration refused to support it. President Bush commented publicly on the Kyoto Protocol and "the important issue of global climate change" on 11 June 2001, a few months after he took office. He said his Cabinet-level working group on global warming "asked the highly respected National Academy of Sciences to provide us the most up-to-date information about what is known and about what is not known on the science of climate change."[2]

He continued: "First, we know the surface temperature of the earth is warming. It has risen by 0.6 degrees C over the past 100 years. There was a warming trend from the 1890s to the 1940s; cooling from the 1940s to the 1970s; and then sharply rising temperatures from the 1970s to today.

"There is a natural greenhouse effect that contributes to warming. . . . Concentrations of greenhouse gases, especially CO_2, have increased substantially since the beginning of the industrial revolution. And the NAS indicates that the increase is due in large part to human activity.

"Yet the Academy's report tells us that we do not know how much effect natural fluctuations in climate may have had on warming. We do not know how much our climate could, or will change in the future. We do not know how fast change will occur, or even how some of our actions could impact it. For example, our useful efforts to reduce sulfur emissions may have actually increased warming because sulfate particles reflect sunlight, bouncing it back into space. And, finally, no one can say with any certainty what constitutes a dangerous level of warming, and therefore what level must be avoided. The policy challenge is to act in a serious and sensible way, given the limits of our knowledge."[3]

As a politician, Mr. Bush has clearly been unwilling to commit the United States to the cost of building an entirely new energy system when (1) the old energy system was still working; (2) the eco-recommended energy systems were expensive and erratic; and (3) the science of global warming was still uncertain.

Also, as a politician, he has been equally unwilling to reject man-made global warming as a myth when many voters believed it was a real danger to the country and the planet.

Russia's Kyoto Minuet

The terms of the Kyoto Protocol required the treaty to sign up countries emitting more than 55 percent of the greenhouse gases before it could go into effect. For Kyoto supporters, the only alternative to U.S. ratification was to sign up Russia.

On 2 December 2003, Russian President Vladimir Putin announced that his country would *not* ratify the Kyoto Protocol. He stated that the treaty was "scientifically flawed," and that "even 100 percent compliance with the Kyoto Protocol won't reverse climate change."[4]

As the *New York Times* reported: "Mr. Putin did not publicly discuss the treaty . . . but [Andrei Illarionov, a senior Putin adviser] stated in a telephone interview that Russia's decision was unequivocal. 'We shall not ratify,' he said.

"Mr. Illarionov said the treaty's supporters had failed to answer questions about the treaty's scientific rationale, its fairness and the potential harm to Russia's economy, which Mr. Putin has pledged to double over the next decade."[5]

The *New York Times* article also noted that even signatory countries were finding it difficult to meet Kyoto limits: "Even as the statements from Russia rocked the treaty talks, the European Commission issued a report warning that the European Union overall, and 13 of its 15 member states, would fail to meet their targets under the Kyoto Protocol unless new measures to curb greenhouse emissions were enacted."[6]

The Russian Academy of Sciences then recommended against Russia's signing Kyoto, following an international seminar on climate change held in Moscow on 5–8 July 2004. The first reason for rejecting Kyoto, the Academy said, is that the world's temperatures do not follow CO_2 levels. It noted that the warmest global climates in the past 2,000 years occurred during the Roman Empire and the Medieval Period, when temperatures were warmer than today—despite lower CO_2 levels. Second, the Russian Academy said there is a much better correlation between world temperatures and solar activity than with CO_2 levels. Third, the Russian scientists noted that world sea levels are not rising faster with warming. Sea levels have been rising about six inches per century since the Little Ice Age ended about 1850. Fourth, the Russian scientists discounted one of the most significant danger claims about global warming: a spread of tropical diseases due to higher temperatures. The Russians noted that malaria is a disease encouraged by sunlit pools of water where mosquitoes can breed, not by climate warmth. Finally, the Russian Academy of Science pointed out the lack of correlation between global warming and extreme weather. Indeed, the British government delegation at the seminar admitted that they could not claim any increase in storms due to climate warming.

Then, however, President Putin changed his mind and the Russian Duma promptly ratified Kyoto. No one is sure why, though Putin had been in negotiation with European governments to get favorable terms for Russian accession to the World Trade Organization. Russia urgently needed to become a member of the WTO, and could gain importantly from being treated as a "developing" economy rather than enrolling as a developed country.

Kyoto's Momentum Decline

What sapped the momentum of the Kyoto Protocol?

For one thing, none of the scary scenarios that had been loudly predicted by the IPCC and various science and environmental groups had actually come closer to reality during Kyoto's seven years of inactive status. No wildlife species extinctions had been tied to warming. Sea levels had not risen faster. Human death rates from warming-related causes had not increased.

During the seven years, the world had also come much closer to 2012 when the second phase of the treaty was supposed to be installed—when its member states would take major steps to stabilize greenhouse emissions to the atmosphere. It was clear to all that the second phase would have to be much more severe than the 5.2 percent cut (from 1990 levels) agreed on for the first phase to 2012.

Perhaps the crucial factor was the surge of industrial growth and wealth in the Third World—and comparatively slow economic growth in Europe. Recently, China's huge economy has been growing more than 8 percent per year and India's at more than 5 percent. The U.S. economy has been growing at 3–4 percent per year. The EU economies have been limping along with twice the unemployment and far lower growth, even without any severe emission cuts imposed under Kyoto.

Some jobs would be created through the Kyoto Protocol in member countries, but far more would be lost through the economic stagnation and the higher taxes that would be required to ration energy use. Many jobs would also be exported from Kyoto member countries to nonmember economies with lower energy costs. Jobs in such energy-intensive industries as mining, metal working, and farming (due to its fertilizer requirements) would be the most likely to be exported. Both Asia and America would benefit at Europe's expense.

It is also true that the entire world watched the California experience: increasing power shortages, rolling blackouts, companies threatening to move themselves and their jobs out of the State, and finally a surge of political demands for the government to license additional power plants, global warming notwithstanding.

THE FINAL FAILURE OF THE KYOTO PROTOCOL

As this book goes to press, no one is sure exactly how the failure of the Kyoto Protocol will play out, but its final failure seems assured.

Russian accession has brought the Kyoto treaty into effect—but the Russians have also brought to the protocol their huge stack of emission credits based on the collapse of the old Soviet industries since 1990. The likely outcome is that Europe will buy these Russian emission credits and make no further real effort to actually reduce CO_2 emissions as the treaty supposedly demands.

Europe's CO_2 emissions were 4,245 million metric tons in 1990 and fell to 4,123 million tons in 2002, thanks to the reductions in coal burning in Britain and the former East Germany. Still, the Kyoto agreement would require the EU to cut its emissions back to 3,906 million tons before 2012. Instead, the EU economies have been emitting *more* greenhouse gas recently. As a result, a report prepared by the UN in December 2003, predicted they would miss their 2012 CO_2 reduction target by 311 million tons.[7]

Russia's CO_2 emissions in 1990 were 2,405 million metric tons and by 2001 they had fallen to 1,614 million tons. Thus the Russians could sell up to 800 million metric tons of emission rights at an "auction" price likely to be far lower than the economic cost of Europe shutting down fossil-fired power plants or forcing trucks off its highways.

European governments will be able to purchase "compliance" with their first-phase Kyoto commitments without actually reducing CO_2 emissions or doubling their already-high gasoline taxes. For its money, Europe will get pieces of paper testifying that CO_2 emissions have been reduced in Russia.

Of the EU countries, only Sweden and Britain have been on target to meet their modest first-phase Kyoto commitments. Moreover, the performance of solar and wind power have been so weak that significant voices in both of those countries are now calling for renewed investments in nuclear power.

Canada is on target to meet little more than half of its Kyoto emissions reduction.[8]

Japan might also buy Russian emission rights. Its Kyoto obligation is a 6 percent *cut* in emissions from 1990 and its current emissions are nearly 8 percent *over* the 1990 levels. The Japanese economy is just emerging from a decade-long recession and a 14 percent cut from current emission levels would likely throw it back into a full-scale depression.

The net effect of the Kyoto Protocol will be to transfer several billion dollars to the Russian government without reducing any real-world CO_2 emissions into the atmosphere.

At Buenos Aires, Italian environment minister Altero Matteoli said, "The first phase of the Protocol ends in 2012; after that it would be unthinkable to go ahead without the United States, China and India. . . . Seeing as these countries do not wish to talk about binding agreements, we must proceed with voluntary accords, bilateral pacts and commercial partnerships."[9]

The governments of the developing countries have no interest in committing to energy constraints. They are extremely conscious that they stay in office only by delivering higher standards of living to their citizens. Television and the Internet have given their people a vision of affluence. Kyoto's energy constraints would destroy their hopes of affluence.

"I think that everybody agrees that Kyoto is really, really hopeless in terms of delivering what the planet needs," Peter Roderick of Friends of the Earth International told CNSNews.com in Buenos Aires. "It's tiny, it's tiny, tiny, it's tiny. It is woefully inadequate, woefully." While Roderick dismissed the potential impact of Kyoto, however, he said the treaty is still "vital" for its symbolic importance.[10]

The life and limbo of the Kyoto Protocol has proved that affluent societies are willing to *talk* about energy rationing without proof that humans are creating a dangerous climate warming. However, Kyoto has also proved that those affluent societies are not willing to actually *impose* energy rationing on their citizens without such proof.

THE ECONOMIC COSTS AND BENEFITS
OF THE 1,500-YEAR CLIMATE CYCLE

The advocates of the Kyoto Treaty have come up with estimates of some enormous costs that will supposedly be inflicted on the planet by man-made warming. However, these cost estimates have shared some serious flaws.

First, the estimates have all been based on radical warming. Little has been said about the economic impact of a 2 degrees Celsius warming, because it is not likely to inflict major costs and should even produce benefits.

Second, the estimates of global warming costs have been inflated in many of these estimates by assumptions of "global warming impacts" that we have already shown in this book to be extremely unlikely or impossible: radical increases in sea levels even though the world has little ice left that can melt rapidly in a moderate warming; higher rates of malaria and other tropical diseases which could be prevented by pesticides, window screens, and other readily-available technologies; assumed crop losses in the tropics that might not occur, and which, in any case, would be outweighed by very large crop yield gains in the big northern cropping regions of Russia and Canada.

Third, the warming-wary ignore some of the known economic benefits of warming, such as (1) the increased crop and forest yields proven to be stimulated by higher levels of CO_2 in the atmosphere, and (2) the lower death rates and reduced need for medical care in a warmer climate as opposed to a colder one.

The huge estimates of global warming costs, in fact, ignore history. The Romans, Chinese, and medieval Europeans all tell us that the last two warming phases of the 1,500-year cycle were prosperous times for humanity. The Roman Empire and the Chinese empire both thrived during the Roman Warming 2,000 years ago. The prosperity of the Medieval Warming is apparent to us today through the beautiful castles and cathedrals of Europe, which date mainly from that period. How could these have been built if the warmings were accompanied by the flooding, epidemics of malaria, massive famine, and constant storms assumed by the gloomy advocates of man-made warming?

Nor do the Kyoto supporters want to look very closely at the costs of giving up fossil fuels. A London economic consulting firm, Lombard Street Research, recently noted that the shift from fossil fuels to whatever low-emission energy systems we adopt would likely cost at least $18 trillion, and perhaps much more.[11] Lombard Street assumed that the shift would take only five years and cost the world half a percentage point of economic growth. It seems obvious that the shift to a totally different energy system is likely to take considerably longer and cost quite a lot more than their admittedly-conservative estimate.

Even so, says Charles Dumas, the Lombard Street lead author, the cost of the warming-prevention strategy is much greater than any conceivable benefit. "This is orders of magnitude greater than the cost of dealing with higher sea levels and freak weather insurance."[12]

One of the more balanced studies of global warming costs and benefits is *The Impact of Climate Change on the U.S. Economy*, written by Robert Mendelsohn of the Yale School of Forestry and James Neumann of Industrialized Economics, Inc. Mendelsohn and Neumann assumed that a doubling of CO_2 in the atmosphere would produce a temperature rise of 2.5 degrees Celsius, and a 7 percent increase in precipitation. They found this would generate large gains in agriculture, and smaller gains in timber and recreation. The other economic sectors would suffer *small* negative impacts. Overall, Mendelsohn and Neumann concluded that the U.S. economy would *gain* slightly from such a warming—by 0.2 percent of gross domestic product (GDP).[13]

This is in sharp contrast to the IPCC's 1995 report, "Economic and Social Dimensions of Climate Change," which had assembled five earlier economic studies. Those studies all estimated sizeable global damages from global warming—but the wide range in estimates by sector indicated there was great uncertainty among the IPPC report's authors. For example, the estimated costs to agriculture ranged from $1.1 billion to $17.5 billion. Estimates of timber losses ranged from $700 million to more than $43 billion—a sixty-fold difference!

Mendelsohn and Neumann estimated that agriculture would gain more than $40 billion from longer growing seasons, fewer frosts, more rainfall, and increased CO_2 fertilization. Agricultural sector studies have indeed shown large gains to farming from global warming. Timber would gain from the same factors. Recreation generally benefits from warmer temperatures. Thus the Mendelsohn-Neumann argument is supported by logic. It also benefits from the authors' inclusion of adaptation strategies and of actual observations on energy expenditures and leisure activities in towns that have experienced temperature changes.

CONCLUSION

Human society should attempt to put binding constraints on human emissions of greenhouse gases only if the advocates of man-made warming can demonstrate three things:

1. That the greenhouse gases are certain to raise global temperatures significantly higher than they rose during previous natural climate warming cycles;
2. That the warming would severely harm human welfare and the ecology;
3. That rational human actions could actually forestall such overheating.

To date, the advocates of man-made warming have not been able to meet any of these minimum requirements. The IPCC's claim to have found the "human fingerprint" in the Modern Warming was bogus when Ben Santer first altered the science chapter in the IPCC's 1996 report, and it remains bogus today.

No one has been able to distinguish natural from man-made warming. Indeed, since the industrialized regions have seen the vast majority of warming, we may be dealing with localized surface heating driven primarily by urban heat islands and land use changes.

GLOBAL WARMING WILL NOT FADE FROM THE FRONT PAGES

Today, we finally have the ice cores, ancient tree rings, and stalagmite analyses to document the 1,500-year climate cycle. We have the satellite readings on the sun's variability. We've documented the atmospheric heat vent over "warm pool" of the Pacific.

If we objectively list the strengths and weaknesses of the two concepts on the same page, the Greenhouse Theory looks woefully weak. The 1,500-year climate cycle looks much more convincing. But man-made "global warming" is in the public mind. A moderate natural 1,500-year climate cycle is not. Moreover, a great many people with a great deal of power have invested heavily in the Greenhouse Theory.

The environmental movement may be a bit past its peak influence but it still gets massive approval ratings in public opinion polls. It is still dedicated to making our society feel guilty about its wealth and materialism and retargeting us toward the leaner society the activists believe we should be forced to accept.

Mainstream journalists have long since committed themselves to the environmental cause. It appeals to their sense of superiority, and it gives them an unending source of scary news for front pages and TV sound bites. How else can the journalists generate front-page bylines in a world where human lives are lengthening, famine is being conquered, and the Cold War's mutually-assured-destruction has disappeared? Even in today's wars (including the war on terror) death totals are reckoned in the thousands, not millions.

The climate research community has become massively dependent on billions of dollars per year in government research grants generated by the global warming campaign. Thousands of new Ph.D.s have been earned, hundreds of new research projects have been undertaken, and dozens of new scientific journals have been founded to publish their climate research results.

If the public was suddenly convinced of the natural, moderate 1,500-year cycle, there would be a crushing impact on the eco-groups' donations and grants, and on the reputations of the journalists who wrote the global warming scare stories, along with professional starvation for various university departments, government laboratories, and whole divisions of NASA and the EPA.

There will certainly be enough "hot" events and enough weather disasters that can be blamed on global warming to keep the subject alive in the media. Beyond that, a sudden surge in global temperature could occur at any time and rekindle fierce public pressure to "do something." In the heat of the moment, could the "something" be a "son of Kyoto"?

A "SON OF KYOTO"?

There now seems very little likelihood that most of the people in the affluent countries will willingly give up their energy-intensive lifestyles. Our homes and workplaces are increasingly climate controlled with fossil-supplied energy. Our societies and stores are organized around the personal mobility provided by cars—with far more fuel-efficient gas-electric hybrids coming over the horizon.

Even our exercise is increasingly taken in air-conditioned buildings. Instead of braving the elements to hike, bicycle, or climb rocks, we have "fitness centers" where we can sweat in convenient comfort—and get into energy-intensive hot showers quickly afterward.

Looking at the California experience with rolling blackouts in 2000, it seems unlikely that democratically elected governments would impose "second stage Kyoto" requirements that would move rapidly to CO_2 stabilization. No American president or senator is likely to endanger their party and/or re-election by moving too rapidly on a long-term problem like human-emitted greenhouse gases. Europe's economics ministers are already feuding with their environmental ministers about how and whether to move on CO_2.

Any "son of Kyoto" is therefore likely to sound tougher than the Kyoto's first stage, perhaps requiring a 25 percent reduction with respect to 1990 emission levels, rather than 5 percent. But instead of carrying a target date of 2012, the next-generation Kyoto might hold off until perhaps 2030 or 2040, well beyond the current politicians' terms in office.

A new climate change treaty would at least pay lip service to the obligations of developing nations, though it could probably not require them to reduce emissions. Instead, the new Kyoto might be shaped by the notion of "contraction and convergence," now popular in European environmental circles.

The concept is that every human being on this planet has the right to emit the same amount of carbon dioxide. Therefore, citizens of developing nations would be given the same quota for emissions as citizens in industrialized nations. The latter would have the privilege of buying unused emission rights from those who are not using their allocated quota. In other words, the world would see a giant cash transfer from developed to developing nations.

The United Nations would be ecstatic. The multibillion-dollar graft of its Iraq Oil-for-Food program would pale in comparison to its becoming the "rationing board" for the world's energy. Would a "son of Kyoto" be any more effective in actually reducing greenhouse gas emissions than Kyoto itself? It's highly unlikely. It would, however, prolong the careers of the greenhouse advocates.

THE REAL COSTS OF PROLONGING CO$_2$ CONSTRAINTS

A "prolong Kyoto" effort would let the global warming professionals continue collecting their billions of dollars per year for studying and scaring. It would also avoid a knock-down fight between the Greens and the man-made warming nonbelievers, who probably outnumber them but have no particular interest in a showdown. This prolonging policy would have major drawbacks, however—even beyond the billions of dollars that would continue to be wasted on nonrealistic climate models.

A "prolong Kyoto" strategy would deny the world the economic growth that could and should be produced if its massive amounts of coal and low-quality hydrocarbon reserves were burned in clean, high tech, high-efficiency systems such as coal gasification. Coal is the most abundant fossil fuel and it now provides about half of America's electric power. The world has about 1,000 billion tons of known reserves according to the Energy Information Agency. That's about two hundred years' worth at recent consumption levels. The United States, Russia, and perhaps the Ukraine have the biggest reserves, but China, Australia, Germany, and South Africa also have large coal deposits—all of them outside the Middle East.

The world also has massive amounts of bitumen, a tar-like mixture of petroleum hydrocarbons, that can be used as power plant fuel when mixed into a slurry with small amounts of water and emulsifier. Venezuela has 270 billion barrels worth of recoverable bitumen reserves, a larger total than Saudi Arabia's proven oil reserves (240 billion barrels). China is already buying it.

Alberta, Canada, says its total reserves of "tar sands" (along with its oil not recoverable by current technologies) may be as high as 2,500 billion barrels. Current world oil consumption is about 30 billion barrels per year.

Oil shale could potentially provide trillions of barrels of a petroleum-like fuel. Today's recovery technologies are expensive, requiring high heat and rock crushing. Could a biological recovery system based on genetically altered bacteria recover that chemical resource at low cost? Would such a breakthrough even be sought in a world trying to avoid CO_2?

If CO_2 is not causing dangerous global warming, there is no reason not to use the coal and tar sands. They should be used carefully and in clean-burn systems to minimize actual air pollution, but there is no virtue to leaving them in the ground for future generations that may have no more use for them than we do for the once-treasured whale oil. In a human society that as yet has no effective replacement for fossil fuels, three hundred years' worth of low-grade fossil fuels could be massively important.

Imagine an America in which you are allowed to drive your auto only two days per month; in which air conditioning is banned from homes and offices; in which the ice man cometh daily, because there are no electric refrigerators. Imagine the Sun Belt being evacuated. How much industrial and business investment would become useless? How many thousands of deaths would we suffer just from food poisoning due to poorly refrigerated food? How many people would die for lack of the high-tech diagnostic equipment currently powered by electricity?

How many millions of subsistence households in the biodiversity "hotspots" of the world will continue to hunt endangered animals for food, clear species-rich and highly erodable lands for low-yield farming, and burn millions of trees they don't replant for cooking and heating? How many poor women will die from indoor air pollution?

ADAPTING TO CLIMATE CHANGE

The warming phase of the 1,500-year climate cycle will bring some moderate changes in global temperatures. Summer temperatures will rise, on average, only slightly. However, there will be uncomfortable heat waves, as there are now.

Millions of Americans have already volunteered for such "global warming" by relocating in the recently air-conditioned South and Southwest. Their climate change is as great as most of the world's inhabitants should expect in the next several centuries, and they love it.

The winter nights will be less cold in the temperate zones, but most people will still need their furnaces.

Close to the equator, rainfall and drought patterns are likely to shift somewhat. There will be droughts in the twenty-first to twenty-third centuries, some of them prolonged. But the physical evidence shows there are always droughts, some of them extensive, whether the climate is warming or cooling.

We cannot protect all regions from climate change disadvantages. We can, however, use our technology to adapt, with less radical changes in our lives than were inflicted on the poor Norse who froze or starved in Greenland, or the Mayans who had to abandon their cities to subsist in drought-stricken jungle. We'll be far better off than the Slavs and Scots driven from their highland farms into landless wandering during the Little Ice Age, when there were virtually no off-farm jobs for anyone.

Food production will change during the climate changes but our high-yield farming will ensure adequate food. We will be readily able to transport food to cities that cannot otherwise sustain themselves. In fact, we already do. We could invest in seawater desalinization for coastal cities such as Los Angeles and San Diego, and improve waste-water recycling for all cities.

Where the really severe droughts appear, we may have to accept the ultimate answer of the Mayans: people may have to move. These days, that's not so hard, especially with government emergency assistance payments. The people who aren't hit by extreme prolonged drought are rich enough to help those who are, as in the Hurricane Katrina disaster in America's Gulf Coast in 2005.

The message from the ice cores is clear: global warming is natural, unstoppable, and not nearly as dangerous as the public hysteria over it.

TRULY HUMAN AND HUMANE ENVIRONMENTALISM

Some years from now a future generation—having survived real threats like international terrorism or weapons of mass destruction—may look back on this episode in human history as a temporary hysteria that briefly gripped much of the Western world. By then, fossil fuels may be mostly depleted, with the cost of energy held in bounds only through massive investments in nuclear power or the development of still-unforeseen energy technologies

The chief worry then is likely to be the coming ice age still looming as our mild interglacial period draws toward a close. The watchwords will then, more than ever, be "insulate," "adapt," and "grow more food on less land," to leave more room for the wildlife pushed toward the equator by the ice sheets.

NOTES

1. Ron Bailey, "The Kyoto Protocol is Dead," 17 December 2004, <techcentral station.com>.

2. "President Bush Discusses Global Climate Change," transcript of remarks by President George W. Bush in the White House Rose Garden, 11 June 2001, <http://www.whitehouse.gov/news/releases/2001/06/11>.

3. "President Bush Discusses Global Climate Change."

4. Steven Lee Myers and Andrew Revkin, "Putin Aide Rules Out Russian Approval of Kyoto Protocol," *New York Times*, 2 December 2003.

5. Myers and Revkin, "Putin Aide Rules Out Russian Approval of Kyoto Protocol."

6. Myers and Revkin, "Putin Aide Rules Out Russian Approval of Kyoto Protocol."

7. Greenhouse gas emission data for all ANNEX I countries can be found at <www.grida.no/db /maps/collection/climate9/index.cfm>.

8. "We'll Fall Short of Kyoto Targets, Ottawa Says," *Toronto Star*, 2 December 2004.

9. Robin Pomeroy, "Italy Calls to End Kyoto Limits," Reuters, 15 December 2004.

10. Marc Morano, "Greens Concede Kyoto Will Not Impact 'Global Warming,'" <CNSNews.com>, 17 December 2004.

11. Brendan Keenan, "Cost of Ending Global Warming 'Too High,'" *Irish Independent*, 18 August 2005.

12. Keenan, "Cost of Ending Global Warming 'Too High.'"

13. Robert Mendelsohn and James Neumann, *The Impact of Climate Change on the U.S. Economy* (Cambridge, UK: Cambridge University Press, 1999).

Glossary

African Sahel—The "edge of the desert," it occupies millions of square miles just south of the Sahara Desert.

Albedo—Meaning "whiteness," albedo is a measure of how much light a surface reflects. Snow and ice have high levels of albedo, while forests and oceans have low albedo.

Algae cyst—Tiny capsule-like sac formed by some plankton for protection under stresses. Fossil remains of these sacs indicate abundance and types of the plankton that lived during the formation of a given layer of sediment

Alien species—Introduced into a region where it is not native, either deliberately or by accident.

Arctic Oscillation—See North Atlantic Oscillation.

Atlantic Conveyor—A massive three-dimensional pattern of ocean currents driven by wind, temperature, and salinity by which the Atlantic Ocean moves heat from the Southern Hemisphere and the equator northward.

Axial tilt—The Earth spins along with an oblique angle between the plane of the equator and the plane's orbit around the sun, which causes our summer and winter seasons.

Biogenic opal—Mostly glassy remains of diatoms, the most common form of plankton.

Biofuels—Fuels made from organic biomass, mainly ethanol (from sugarcane or corn), and biodiesel (from soy or other vegetable oils). Methanol was originally made from wood, but is now mostly produced from natural gas.

Biological pump—A major element of the planet's carbon cycle. Trillions of tiny plankton in the oceans remove carbon from the atmosphere. As they and the fish that eat them die, their detritus (containing billions of tons of carbon) sinks to the ocean floor.

Boreholes—Historic temperatures lingering in deep layers of rock and soil can be recovered by measuring their downward transmission at various depths. Borehole readings become too faint to recover beyond 1,000 years or so.

Bristlecone pines—The world's longest-lived trees, living up to 5,000 years in the short growing seasons and dry, rocky soils of the upper treelines in western America.

Bubonic plague—The famous "black plague" that repeatedly struck Medieval Europe. It is primarily a disease of rodents, with fleas as intermediate hosts. Outbreaks often occur when other rodent diseases or food shortages have destroyed the fleas' normal rodent hosts.

Carbon cycle—Animals breathe in oxygen and exhale carbon dioxide. Plants take in carbon dioxide and release oxygen. Massive amounts of carbon are stored, primarily as limestone and dolomite, which have been formed by such living organisms as algae and corals. Additional carbon is stored in seawater and soils. Less than 2 percent of the carbon is held in the atmosphere.

Carbon dioxide—The fourth most abundant gas in the atmosphere. It makes up only 0.038 percent of the total gases, though that is up from 0.028 percent in the preindustrial era. CO_2 is created by the decomposition of organic matter and by combustion. Animals exhale it, and plants convert it into growth.

Carbon dioxide fertilization—The growth of most trees and plants is stimulated by higher concentrations of CO_2 in the atmosphere. Water use efficiency is also improved. The fertilization effect has been documented in a wide variety of trees and plants

Cariaco Basin—A broad, deep geologically stable basin on the floor of the Caribbean, a natural sediment trap, that reveals much about past climate processes.

Cirrus clouds—High, wispy clouds, generally above 20,000 feet, composed mainly of ice crystals. They tend to seal in atmospheric heat,

Climate proxy—Physical evidence that indirectly indicates past climate conditions, such as ice cores, fossils, tree rings, corals, and oxygen isotopes in fossilized seashells.

Closed forest—A mature forest in which the crowns of the existing trees close off most direct sunlight.

Cloud chamber—A closed container filled with supersaturated water vapor. When radiation passes through the vapor, the particles leave distinctive trails of mist.

Coal gasification—Coal is partially oxidized with steam and oxygen, and the resulting gas is scrubbed of sulfur and particles before being burned in a gas turbine. The CO_2 comes off in a dense stream that is easy to capture.

Comet—Chunks of frozen gases, ice, and rocky debris that orbit the sun. Heat from the sun can produce a "tail" millions of miles long. The galaxy probably has trillions of comets, but few come close to Earth

Corals—The external skeletons of tiny marine anthozoans built from calcium carbonate, which contains oxygen and its isotopes, as well as trace metals.

Cosmic rays—High-energy particles that originate in the galaxy beyond Earth, and play a significant role in the natural mutation and evolution of life forms on Earth. They consist mostly of the nuclei of atoms, which collide at nearly the speed of light with other atomic nuclei in the upper atmosphere

Cryosphere—Any part of the Earth's system made up of frozen water: snow, ice sheets, glaciers, permafrost, or floating ice. All are directly related to sea level change.

Cumulus clouds—Low, wet clouds that look like puffs of cotton in the sky. They tend to cool the Earth by reflecting solar heat back into space.

Cyclone (tropical)—The worst storms in the Indian and western Pacific Oceans. They are called hurricanes in the Atlantic and typhoons in the northwest Pacific. They have sustained winds of at least sixty miles per hour, can dump up to three feet of water within twenty-four hours and attack coastlines with thirty-foot storm surges.

Dansgaard-Oeschger cycle—The formal name for the irregular 1,500-year climate cycle, discovered by Denmark's Willi Dansgaard and Switzerland's Hans Oeschger in the Greenland ice cores in 1983.

Dark Ages—the climate time period following the Roman Warming that began about A.D. 600. The Earth's climate turned cold and unstable, with major droughts across Eastern Europe, the Middle East, and Asia.

Diatoms—One-celled algae that are the most common form of plankton. A favorite research tool because their silica walls are often well preserved in sediments.

Dinoflagellate cysts—The "skeletal remains" of tiny marine creatures, mostly protozoans, typically abundant in marine sediments.

El Niño/Southern Oscillation (ENSO)—A huge periodic shift in ocean currents and atmospheric patterns that seesaws between the eastern Pacific and the western Pacific/Indian Ocean. The ENSO Index is the difference in air pressure between Tahiti and Darwin, Australia. The ENSO events occur every three to five years and last twelve to eighteen months.

Emission credits—Under the Kyoto Protocol, countries emitting less than their allotments of greenhouse gases receive "emission credits," which could be sold to other countries that have failed to meet their emission reduction targets.

Extinction—The complete disappearance of a species; the death of the last individual of that species. Millions of extinctions have occurred in geologic time, but most have been associated with catastrophic events, such as asteroid impacts that blocked out the sun for years at a time and thus created massive, sudden changes in Earth's climate.

Faunal record—Our knowledge of the Earth's animal life revealed primarily in fossils.

"Fingerprinting" climate change—The Intergovernmental Panel on Climate Change in its 1995 and 2001 reports suggested that anthropogenic climate warming could be deduced from correlations between observed and modeled patterns of climate-related variability. It is still a matter of controversy.

Framework Convention on Climate Change—The UN treaty on climate change signed by more than 150 countries at the Rio Earth Summit in 1992. Its stated goal is to stabilize the levels of greenhouse gases in the Earth's atmosphere at levels that will not represent a danger to the "climate system." The Kyoto Protocol is an outgrowth of this treaty.

Fusion reactor—The sun is a giant fusion reactor, and scientists are striving to create a miniature version of it to power human society in the latter twenty-first century. In fusion, several atomic nuclei reform into a larger nucleus, which has slightly less mass than the original nuclei combined. The small amount of lost mass turns into very large amounts of energy, as the sun demonstrates

General Circulation Models (GCMs)—Huge computer models that attempt to simulate the climate forces on Earth, usually in response to forcing by greenhouse gases. The models have thus far not been able to explain past world temperatures, nor have they been validated by "ground truth" data.

Geologic record—The millions of years since the Earth was formed, recorded in the physical evidence of rocks, fossils and landforms.

Gigatonne—One thousand million tons (one metric ton is 1,000 kilograms, or 2,200 pounds).

Glacier—A huge mass of ice flowing slowly overland in a region where annual precipitation exceeds melting.

Glacier advances and retreats—Good indicators of longer-term trends because they reflect average climate conditions over decades or longer. Piles of glacial till left behind during retreats clearly mark the glaciers' maximums, and organic matter in the tills can be carbon-dated.

Greenhouse gases—Any of the gases that trap heat radiation in the Earth's atmosphere. The most important is water vapor, which contributes 36–70 percent of the atmosphere's natural greenhouse effect. CO_2 contributes 9–26 percent, ozone 3–7 percent. Lesser greenhouse gases include methane, nitrous oxide, and chlorofluorocarbons.

Greenhouse Theory—The theory that human emissions of greenhouse gases, especially CO_2, are heating the Earth beyond the levels of past climate variation, creating serious danger for the planet's ecosystems.

High-altitude weather balloons—Since the 1960s, the world's meteorological services have sent aloft large balloons carrying radiosondes that relay back atmospheric pressure, temperature at various altitudes, and humidity. Radar monitoring of the balloons' movements provides information on winds.

Holocene—The warm interglacial period since the last ice age ended 11,000 years ago.

Holocene Climate Optimum—The longest and warmest "warming" of the present interglacial, which occurred from 9,000 to 5,000 years ago. Greenland ice cores show temperatures then were 2.5 degrees Celsius warmer than present, for 3,000 years.

Huon pines—Among the slowest-growing and longest-lived tree species in the world, with individual specimens living to 3,000 years or more. They are found on Tasmania's west coast and in its Huon Valley. The huon pine is not actually a pine but a podocarp (Dacrydium franklinii), a relic from 135 million years ago.

Ice ages—They have occurred on Earth about every 100,000 years, each lasting about 90,000 years. During the ice ages, the Northern Hemisphere is covered with huge ice sheets, sea levels are hundreds of feet lower, and temperatures are 7–12 degrees Celsius lower than during the interglacials. The Milankovich Theory is that ice ages are caused by periodic changes in the Earth's orbit around the sun.

Ice sheet—Extensive sheets of ice that cover large areas, such as in Greenland and the Antarctic, where they accumulate thousands and even millions of years of precipitation.

Ice wedges—Vertical sheets of ground ice that form in thermal cracks of permafrost when temperatures are at least –40°C. They can "grow" from year to year as more water seeps into the permafrost and then freezes into the wedge.

Infrared light/heat—Has a longer wavelength than the visible light spectrum and thus cannot be seen with the naked eye.

Interglacial periods—The periods between ice ages, usually about 10,000 years in length. Our modern interglacial seems to have begun about 12,000 years ago. We cannot really predict when it will end.

Intergovernmental Panel on Climate Change (IPCC)—The advisory body established by the United Nations' World Meteorological Organization (WMO) and Environmental Programme (UNEP).

Intertropical Convergence Zone (ITCZ)—The belt of trade winds, clouds and doldrums that surround the Earth at the equator. The hot sun and warm water produce rising airflow, lots of evaporated moisture, and massive, nearly constant thunderstorms. The ITCZ shifts north and south with the seasonal location of the sun. Longer-term shifts in the zone can drastically change rainfall in equatorial nations, including major droughts and floods.

Isotopes—Most elements have several forms, or isotopes, which differ in the number of neutrons in their nuclei. The heavier isotopes react more slowly than their lighter cousins. Isotopes are useful tracers of climate processes.

Kyoto Protocol—An international treaty negotiated at Kyoto, Japan, in 1997. It obligated member nations to slightly reduce their emissions of greenhouse gases. The U.S. Senate has never ratified the treaty.

Land use change—Any land use other than the mature forests and wild meadows that characterized primeval wilderness.

Laurentide ice sheet—A huge ice sheet that spread over much of Canada and the northern United States during the last ice age. Its five million square miles included parts of Iowa, Illinois, Indiana, and Ohio.

Lichens—Crust-like growths on rocks and tree trunks, formed by fungi growing symbiotically with green algae.

Little Ice Age—The period of cold, unstable climate that abruptly followed the Medieval Warming from 1300 to 1850. Famines occurred frequently, due to the shorter growing seasons, unusual frosts and cloudy skies.

Lower atmosphere (*troposphere*)—The lowest layer of the atmosphere, which extends up to the tropopause (16 kilometers or 50,000 feet in the tropics). Its temperatures decline with altitude—while those in the upper atmosphere increase with altitude.

Mass spectrometer—A scientific instrument that identifies chemicals and isotopes though their unique molecular weights. Often this is done by turning samples into gases at high temperatures (above 400 degrees Celsius).

Medieval Warming—The warm, stable half of the most recent 1,500-year climate cycle, which began in Europe about 900 and persisted there until after 1300.

Metanalysis—A study that combines all available studies on a subject, an overview of research.

Methane—CH_4 is an abundant greenhouse gas whose concentration in the atmosphere has doubled in the past 150 years. It is the principle component of natural gas, and is also given off by animal wastes and rice paddies.

Midges—Tiny insects with a narrow temperature tolerance, often preserved in sediments, which can produce temperature histories accurate to within 1 degree Celsius.

Milky Way—The galaxy of which the Earth and sun are part. It is 100,000 light-years across, and 10,000 light-years thick, and has six spiral arms winding out from the nucleus like a giant pinwheel.

Monsoon—A major Asian wind system that brings summer rainfall from the south or southwest.

Moraines—Deposits of rocks and glacial debris left behind when a glacier retreats.

Nitrogen fertilizer—N is the most important plant nutrient crop plants take from the soil as they grow. Nitrogen fertilizer is taken from the air—which is 78 percent N_2 through an industrial process requiring heat and pressure, usually using natural gas.

Nongovernmental organizations (NGOs)—Groups that have proliferated in recent years, some of which take a role in the environmental regulatory process. They are usually nonprofit, though collectively they receive billions of dollars per year from memberships, subscriptions, and foundation grants

North Atlantic Oscillation (NAO) *or Arctic Oscillation*—Dominates wintertime temperatures and precipitation across the North Atlantic, creating much of the variability in the region's temperatures. Normally, sea-level atmospheric pressures are low in the North Atlantic when they are high in the subtropical Azores in the mid-Atlantic. This drives the "normal" surface winds and storms from west to east across the North Atlantic, bringing warm, moist air to Europe. When the NAO reverses itself, winters are often harsher in northern Europe and weather is wetter in central and southern Europe and the Mediterranean. For most of the twentieth century, the NAO alternated between its positive and negative phases. Since the 1970s, however, it has tended to stay in its positive phase, bringing relatively high temperatures to the United States and northern Eurasia.

Ordovician Period—A time period that began 490 million years ago and lasted 50 to 80 million years. There was one continent, Gondwana, and a rich diversity of marine life. Though temperatures were much like those of today, CO_2 levels were up to 18 times higher than currently. In spite of the high CO_2 levels, an ice age occurred toward the end of the period.

Ozone—O_3 is composed of three oxygen atoms. In the stratosphere, where it occurs naturally, it helps protect the Earth from harmful UV radiation. In the lower atmosphere, it is simultaneously a chemical oxidant, a greenhouse gas, and a major component of smog.

Pacific Decadal Oscillation (PDO)—A newly identified and long-lived climate cycle of the Northern Pacific. Its events seem to last twenty to thirty years, much longer than the yearlong El Niño/Southern Oscillation (ENSO) cycles of the southern Pacific. The oscillation is most visible off the Pacific Northwest coast, where it was first noted in the pattern of Alaskan and Oregon-Washington salmon runs. "Cool" PDO phases favor salmon in the Columbia River, while warm phases send the salmon nutrients to the Gulf of Alaska instead. Farther south, the cool phases favor anchovies and warm phases favor sardines.

Paleoclimate—The study of climate changes since the Earth was formed.

Perihelion—The point of the Earth's orbit where it is closest to the sun. At present, this occurs in January.

Permafrost—Permanently frozen soil or subsoil, characteristic of the Arctic, subarctic, and very high mountain slopes. Permafrost can be as deep as 450 meters.

Phytoplankton—Floating microplants, mainly algae, which live in the sunlit layers of the ocean. They are one of the most basic elements in the marine food chain.

Pikas—Small, furry, tailless rodents that live above the tree lines in North American and Eurasian mountains. They eat green vegetation and also cut, dry, and store it for winter food.

Pollen—Fine, powder-like grains of fertilizing material produced by trees and plants. Pollen provides one of the most accurate records of past ecologies, because each species' pollen has a unique shape, and most pollen falls within a mile of where its plant source grew. It can survive virtually unaltered for millions of years under water, in sediments or peat bogs, and in layers of rock. Pollen tells us, for example, that palm trees and tree ferns extended into Wyoming and Montana 50 million years ago.

Precession—A periodic motion of the Earth's orbital axis, produced by the gravitational pull of the sun and moon on the bulge in the Earth's mid-section near the equator. This slow, conical motion around the orbital axis takes 26,000 years.

Radiocarbon dating—A dating method based on the radioactive decay rate of carbon-14 in organic material. Carbon-14 is produced by cosmic rays hitting the Earth's upper atmosphere.

Range—The extent of habitat where a species successfully sustains itself.

Renewables—Power sources that are not "used up" as we generate power, such as solar, wind, wood, and water power.

Roman Warming—The warm half of the climate cycle that also included the cold Dark Ages. The Roman Warming lasted from about 200 B.C. to A.D. 600.

Sagas—Ancient Scandinavian oral histories.

Sediments—Deposited in annual layers on sea and lakes floors, sediment contains pollen grains, minerals, and the remains of microflora and tiny single-celled animals whose shells can be carbon-dated. Iron, from wind-blown dust, can indicate rainfall regimes.

Soil erosion—Soil movement due to wind and/or water. Topsoil erosion can reduce land productivity.

Solar Irradiance—The amount of electromagnetic energy (ultraviolet, visible, and infrared) arriving at a surface in a specific time period. The Earth's average is about 1,367 watts per square meter (the "solar constant") though we now know that the total varies by about 0.1 percent during solar cycles that last eight to fourteen years.

Solar power—Common prescription for replacing fossil fuels. Solar installations can range from south-facing windows that catch winter sun to large reflective concentrators that trap heat and to photovoltaic cells that turn sunlight into electricity.

Solar wind—A stream of ionized gases, mostly hydrogen, that radiates out from the sun at a million degrees F and speeds of up to two million miles per hour. The stronger the solar wind, the more it shields the Earth from cosmic rays.

Species—Plants or animals belong to the same species if they can interbreed to produce fertile offspring. If the interbreeding test cannot be evaluated, organisms are put in the same species if they have adequate anatomical similarities.

Stalagmites—Cone-shaped deposits formed on the floor (called *Stalactites* when hanging from the cave ceiling) of caves by mineral-rich water dripping down. The mineral content of the stalagmites varies with climate conditions, and they contain isotopes of oxygen and other elements which vary with temperature.

Storm surge—Huge coastal waves produced by the low atmospheric pressure and high winds around the eye of a hurricane or typhoon. They are often more destructive than the high winds.

Stratosphere—See Upper atmosphere.

Strandplains—Shore-parallel beach ridges that define ancient seashores and sea levels. They often contain organic material that can be carbon-dated.

Sulfate aerosols—Atmospheric particles produced by the burning of fossil fuels, especially coal. When global warming failed to increase as much as the early model forecast had projected, the Intergovernmental Panel on Climate Change suggested that sulfate aerosols had deflected incoming light, producing a cooling effect and offsetting the effect of increased CO_2. Research, however, demonstrated that regions producing lots of sulfate aerosols tended to be warmer than those which did not.

Sunspot Minimum—Humans have been monitoring sunspots for more than 2,000 years. The overall impact of fewer sunspots is cooling of the Earth.

- The Sporer sunspot minimum occurred from 1460 to 1550, just before the coldest period of the Little Ice Age began. (Sporer was a German astronomer.)
- The Maunder Minimum, named for an English astronomer, occurred between 1645 and 1710 just before the second ultra-cold phase of the Little Ice Age.

- The length of the sunspot cycles (they range from eight to fourteen years) is an even more accurate guide to the sun's warming than the number of sunspots, with a longer cycle bringing more warming.

Sunspots and faculae—Sunspots are planet-sized dark spots on the face of the sun, where the gases are cooler. Faculae are harder-to-see hotter spots on the face of the sun, also associated with sunspots. The heat of the faculae more than offsets the cooler sunspots, so lots of sunspots mean a warmer sum.

Tidal estuary—A semi-enclosed riverbed, sound or bay, where salt water mixes with fresh water as the tides rise and fall.

Thermohaline circulation—See Atlantic conveyor.

Titanium—Metallic, corrosion-resistant element widely distributed in Nature. In sediments, low concentrations of titanium are often associated with low levels of rainfall.

Tornado—A localized, violent windstorm over land, with a funnel-shaped cloud extending toward the ground.

Tree rings—The annual growth rings formed in tree trunks, which reflect growth conditions year by year. They are wider during good growing seasons (warmer, wetter) than during bad seasons (cold, arid). Tree rings can precisely date droughts, and even record such climate-altering events as major volcanic eruptions. Tree ring studies can go farther back into time with long-dead tree trunks and stumps that have been buried in peat bogs and glaciers, or kept underwater by refilled lakes. Researchers have even cored structural timbers from ancient buildings.

Troposphere—See Lower atmosphere.

Tundra—Treeless land between ice regions and tree lines in the Arctic and subarctic. The soil is black and mucky and the subsoil permanently frozen. Vegetation is dominated by mosses, lichens, and low shrubs.

Typhoon—A hurricane in Asia or the western Pacific.

Ultraviolet radiation—Short-wavelength electromagnetic radiation (less than 400 nanometers) emitted by the sun. It is invisible to the naked eye. UV-A rays induce tanned skins in humans. UVU-B rays, with a shorter wavelength, induce sunburn and are associated with most skin cancers.

Upper atmosphere—The stratosphere, from about 30,000 feet to 55,000 feet, which contains the ozone layer. Temperatures here rise with altitude, instead of declining with altitude as in the troposphere, the layer below it.

Upwellings—Ocean locations, where cool, nutrient-rich waters from the depths rise to the surface because prevailing winds push the warm surface waters away. Upwellings often create rich fisheries, such as the anchovy/sardine fishery off the west coast of Peru.

Urban heat island—Higher temperatures are characteristic of cities and towns because roads, buildings and even lawns tend to absorb more heat than forests and fields.

Warm Pool of the Pacific—A region about the size of North America in the western Pacific, which maintains a virtually constant temperature of 29°C or 85°F. Warm water is fed into the region by the trade winds, while heat and moisture pass from it up into the atmosphere, feeding many storms. Periodically, the region shifts its pattern to produce El Niño/Southern Oscillations.

Weather satellites—Since 1979, satellites in polar and geostationary orbits have been collecting enormous amounts of real-time climate information, including cloud formation, cloud cover, water vapor at various altitudes, land temperatures, sea surface temperature, winds, solar energy being reflected by various parts of the Earth, heat energy leaving the Earth, and precipitation. Satellites are collecting the first accurate sea surface readings from the 70 percent of the planet covered by water.

Wind turbines—Slow-turning windmills which are relatively safe for birds (although perhaps not for bats) but quite costly per watt of electricity produced.

World Meteorological Organization (WMO)—A United Nations agency and a technical center for weather data.

Younger Dryas Event—A sudden, short-lived cooling event that occurred 12,800 years ago during the recovery from the last Ice Age. A surge of warming began melting the huge mile-thick ice sheets that then covered most of North America and Northern Europe. Massive amounts of heavy fresh water overwhelmed the Great Atlantic conveyor's saltwater convection currents. The currents stopped bringing up warmth from the south. The ice cores tell us that in less than 50 years, the atmosphere in North America cooled again by 5–6 degrees Celsius for another 1,300 years. The rest of the Earth was apparently still warming and when the northern glaciers began to melt again, the Greenland warming may have taken only twenty years.

Zooplankton—Tiny animals that drift in sunlit waters and feed on smaller phytoplankton. The smallest are foraminifera, which are widely found in sea sediments and useful in analyzing past Earth climatic shifts.

Index

About the Authors

S. Fred Singer, climate physicist, is internationally known for his work on climate, energy, and environmental issues. He is professor emeritus of environmental science at the University of Virginia, and currently serves as Distinguished Research Professor at George Mason University. He is also president of The Science & Environmental Policy Project, a nonprofit policy research group he founded in 1990.

Dr. Singer was the founding dean of the School of Environmental and Planetary Sciences at the University of Miami, first director of the U.S. National Weather Satellite Service, and served five years as vice chairman of the U.S. National Advisory Committee on Oceans and Atmospheres.

He is the author or editor of more than a dozen books and monographs, including *Global Climate Change* (1989), *Hot Talk, Cold Science: Global Warming's Unfinished Debate* (1997), and *Climate Policy—From Rio to Kyoto* (2002).

Dr. Singer has also published more than four hundred technical papers in scientific, economic, and public policy journals. Most recently, he has coauthored a series of peer-reviewed papers on the important and growing disparity between the world's official thermometers and the above-ground temperature measurements from high-altitude weather balloons and satellites.

Dr. Singer did his undergraduate studies at Ohio State University and earned his Ph.D. in physics from Princeton University.

He lives in Crystal City near Arlington, Virginia.

Dennis T. Avery has been a senior fellow of the Hudson Institute since 1989. Prior to that, he was a senior analyst in the U.S. Department of State (1980–1988), where he won the National Intelligence Medal of Achievement in 1983. He also holds outstanding performance awards from the U.S. Department of Agriculture and the U.S. Commodity Futures Trading Commission.

Avery currently writes a weekly column on environmental issues which is distributed to newspapers throughout the country. His writings have appeared in the *Wall Street Journal*, *Christian Science Monitor*, *St. Louis Post-Dispatch*, *Miami Herald*, *Seattle Times*, *Des Moines Register*, and dozens of other newspapers. He has also been featured in *Fortune*, *Forbes*, *The National Journal*, and *The Atlantic Monthly* ("Will Frankenfoods Save the Planet?" [October 2003]).

Avery's book S*aving the Planet with Pesticides and Plastic: The Environmental Triumph of High-Yield Farming* was first published in 1995, with a second edition in 2000.

The inspiration for this book came when Avery wrote a 1998 article in Hudson's *American Outlook* magazine titled "Global Warming: Boon for Mankind?" The article noted that "the Medieval Warming was one of the most favorable periods in human history. Crops were plentiful, death rates diminished, and trade and industry expanded—while art and architecture flourished." The article was later condensed in *The Reader's Digest* (August 1999).

He lives on a small farm in the Shenandoah Valley of Virginia with his wife, Anne.